Agricultural Finance and Opportunities for Investment and Expansion

Augustine O. Ejiogu
Imo State University, Nigeria

A volume in the Advances in Finance, Accounting, and Economics (AFAE) Book Series

IGI Global
DISSEMINATOR OF KNOWLEDGE

Published in the United States of America by
IGI Global
Business Science Reference (an imprint of IGI Global)
701 E. Chocolate Avenue
Hershey PA, USA 17033
Tel: 717-533-8845
Fax: 717-533-8661
E-mail: cust@igi-global.com
Web site: http://www.igi-global.com

Library of Congress Cataloging-in-Publication Data

Names: Ejiogu, Augustine Odinakachukwu, 1961- author.
Title: Agricultural finance and opportunities for investment and expansion / by Augustine Odinakachukwu Ejiogu.
Description: Hershey, PA : Business Science Reference, [2018]
Identifiers: LCCN 2017012418| ISBN 9781522530596 (hardcover) | ISBN 9781522530602 (ebook)
Subjects: LCSH: Agriculture--Finance. | Agriculture--Economic aspects.
Classification: LCC HD1437 .E395 2018 | DDC 630.68/1--dc23 LC record available at https://lccn.loc.gov/2017012418

This book is published in the IGI Global book series Advances in Finance, Accounting, and Economics (AFAE) (ISSN: 2327-5677; eISSN: 2327-5685)

British Cataloguing in Publication Data
A Cataloguing in Publication record for this book is available from the British Library.

Advances in Finance, Accounting, and Economics (AFAE) Book Series

ISSN:2327-5677
EISSN:2327-5685

MISSION

In our changing economic and business environment, it is important to consider the financial changes occurring internationally as well as within individual organizations and business environments. Understanding these changes as well as the factors that influence them is crucial in preparing for our financial future and ensuring economic sustainability and growth.

The **Advances in Finance, Accounting, and Economics (AFAE)** book series aims to publish comprehensive and informative titles in all areas of economics and economic theory, finance, and accounting to assist in advancing the available knowledge and providing for further research development in these dynamic fields.

COVERAGE

- Evidence-Based Studies
- Auditing
- Economics Geography
- Economics of Migration and Spatial Mobility
- Field Research
- Environmental and Social Accounting
- Economic Policy
- Wages and Employment
- Economics of Risks, Uncertainty, Ambiguity, and Insurance
- Economics of Agriculture and Biotechnology

IGI Global is currently accepting manuscripts for publication within this series. To submit a proposal for a volume in this series, please contact our Acquisition Editors at Acquisitions@igi-global.com or visit: http://www.igi-global.com/publish/.

The Advances in Finance, Accounting, and Economics (AFAE) Book Series (ISSN 2327-5677) is published by IGI Global, 701 E. Chocolate Avenue, Hershey, PA 17033-1240, USA, www.igi-global.com. This series is composed of titles available for purchase individually; each title is edited to be contextually exclusive from any other title within the series. For pricing and ordering information please visit http://www.igi-global.com/book-series/advances-finance-accounting-economics/73685. Postmaster: Send all address changes to above address.

Titles in this Series

For a list of additional titles in this series, please visit:
https://www.igi-global.com/book-series/advances-finance-accounting-economics/73685

Uncertainties and Risk Assessment in Trade Relations
Marianne Ojo (George Mason University, USA)
Business Science Reference • ©2018 • 283pp • H/C (ISBN: 9781522541318) • US $225.00

Economic Reforms for Global Competitiveness
Denis Ushakov (International College Suan Sunandha Rajabhat University, Thailand & Russian Academy of National Economy and Public Administration–South, Russia)
Business Science Reference • ©2018 • 363pp • H/C (ISBN: 9781522538561) • US $235.00

Alternative Decision-Making Models for Financial Portfolio Management...
Narela Spaseski (International University of Sarajevo, Bosnia and Herzegovina)
Business Science Reference • ©2018 • 333pp • H/C (ISBN: 9781522532590) • US $175.00

Ranking Economic Performance and Efficiency in the Global Market Emerging Research...
Benjamen Franklen Gussen (University of Southern Queensland, Australia & Australian Law and Economics Association, Australia)
Business Science Reference • ©2018 • 372pp • H/C (ISBN: 9781522527565) • US $195.00

Financial Entrepreneurship for Economic Growth in Emerging Nations
Atsede Woldie (University of South Wales, UK) and Brychan Thomas (University of South Wales, UK)
Business Science Reference • ©2018 • 304pp • H/C (ISBN: 9781522527008) • US $225.00

Technological Integration as a Catalyst for Industrial Development and Economic Growth
Bryan Christiansen (PryMarke, LLC, USA) and Ülkü Yüksel (The University of Sydney Business School, Australia)
Business Science Reference • ©2017 • 272pp • H/C (ISBN: 9781522523192) • US $185.00

Handbook of Research on Economic, Financial, and Industrial Impacts on Infrastructure...
Ramesh Chandra Das (Katwa College, India)
Business Science Reference • ©2017 • 466pp • H/C (ISBN: 9781522523611) • US $230.00

For an entire list of titles in this series, please visit:
https://www.igi-global.com/book-series/advances-finance-accounting-economics/73685

IGI Global
DISSEMINATOR OF KNOWLEDGE

701 East Chocolate Avenue, Hershey, PA 17033, USA
Tel: 717-533-8845 x100 • Fax: 717-533-8661
E-Mail: cust@igi-global.com • www.igi-global.com

Table of Contents

Preface

This book is about sources and profitable uses of funds in agricultural enterprises. It seeks to highlight the overarching importance of agriculture in the economy and consequently the need to treat agriculture as a business and thereby harness the enormous potentials with a view to fully realizing them. The book advances the idea that to achieve sustainable increases in agricultural production, a public private partnership arrangement should aim at building resilient markets, resilient agriculture and resilient people.

According to Zhang (2017), China will set a deadline for automakers to end sales of fossil-fuel-powered vehicles; thus, China becomes the biggest market to do so in a move that will accelerate the push into the electric car market. Although the move is aimed at impacting on environment and growth of China's auto industry, it sends strong signals to countries whose economies depend on revenues from fossil fuel. Nigeria is one of those countries whose economies depend on revenues from fossil fuel.

This emphasis on agriculture is timely and topical since China, the world's second-biggest economy which is committed to cap its carbon emissions by 2030 and curb air pollution is the latest to join countries such as the UK and France in seeking to phase out vehicles using gasoline. According to the editorial of a Nigerian newspaper, the Independent, of 21st September 2017, in July 2017, the UK is set to ban sale of diesel and gasoline fueled cars by 2040, two weeks after, France announced a similar plan to reduce air pollution and meet targets to keep global warming below 2^0Celcius (3.6^0 Fahrenheit). Norway and Netherlands are considering a more aggressive way to put an end to fossil fuel cars years earlier than its European peers.

According to Zhang, the implementation of the ban for such a big market as China cannot be later than 2040. According to the Independent newspaper editorial of 21st September 2017, the planned ban by China is to be operative by 2025. This book therefore comes at a time when the emphasis of Nigeria government and other governments that solely rely on the revenue from oil to focus on diversification to such areas as agriculture.

The focus on the overarching importance of agriculture in the economy and consequently on the need to treat agriculture as a business with a view to harnessing and realizing the enormous potential is both timely and topical. There is the need to search for alternative sources of revenue other than oil which tends to exhibit wild fluctuations.

The methodology is descriptive and prescriptive. The theoretical underpinnings are based on an application of the theory of the vicious circle of poverty in which low income, gives rise to low savings, low investment and ultimately low productivity, and on the need for external injection of capital into the agricultural activities of the small-scale farmers for the vicious circle to be broken. The book contains the various components of the external capital injection required to make agricultural enterprise a going concern.

In Chapter 1, under the role of agriculture in the development process, we operationally defined agriculture, conceptualized the terms "development" and "economic development", ex-rayed the expected roles of agriculture in the development process, examined the performance of the agriculture sector, examined the role of agricultural credit in agricultural development, examined the concept and measurement of extreme poverty.

In Chapter 2, which focused on the concept of finance for agriculture, we identified and described the three broad sources of accessing funds for agricultural purposes. We further highlighted the three important functions of an enterprise, discussed the meaning of agricultural finance, and described the sources of investment fund for agriculture.

Chapter 3 focused on the economics of farm management by specifically defining the meaning, the scope, and highlighting the resources available to the farm enterprise. On the basis of encouraging agri-business and supporting the large population of smallholder farmers in the face climate change, Chapter 4 highlighted the financing and training imperatives for resilient agriculture in Nigeria.

Chapter 5 particularly focused on engaging in agriculture with a view to successfully growing it as a business. The chapter discussed enterprise expansion and opportunities for expansion in agriculture. Chapter 6 analyzed the anchor borrowers' programme of the government with a view to drawing more attention to its strengths and weaknesses. Chapter 7 focused on the concept of financial analyses by describing the tools for analyses and showing worked examples. In Chapter 8, the two broad categories of measures of project worth namely discounted measures of project worth and undiscounted measures of project worth were presented. Discounted measures of project work take into consideration the time value of money while undiscounted measures do not. In Chapter 9, under farm planning and analysis, the need for planning and analysis of farm enterprises to keep track of profitability and expansion of the enterprise was emphasized. The chapter focused on the basic ways of planning an investment using the instrument of budget. Chapter 10 focused on the costs associated with borrowing and the procedures for calculating interest charges. It also presented the difference between cost of capital and cost of funds. Chapter 11 highlighted the concepts of valuation and depreciation of farm assets. It discussed different methods of calculating depreciation and offered useful insight into how to value asset that appreciates in value. Chapter 12 investigated the concept of agricultural risk management and insurance

REFERENCES

Zhang, Y. (2017). *China Fossil Fuel Deadline Shifts Focus to Electric Car Race*. Bloomberg News.

Chapter 1
The Role of Agriculture in the Development Process

ABSTRACT

Agriculture plays the role of providing employment, income, food, raw materials, and foreign exchange earnings for people. The ability and the inability of agriculture in playing the provisioning roles, in varying degrees, define the poverty status of those engaged in it. It is a paradox that a majority of those who are engaged in agriculture, especially in developing countries, tend to be associated with such poverty-linked characteristics as low income, hunger, deprivation, and vulnerabilities. There is therefore the need to refocus on defining the concept of agriculture with a view to bringing out its role in the development process and how the roles can be effectively achieved by the majority of those engaged in it. The objectives of the chapter include describing the expected roles of agriculture in the development process; highlighting the performance of the agriculture sector; describing the role of agricultural credit in agricultural development; defining the concept of extreme poverty; highlighting some of the strength and weaknesses of incometrics, highlighting vulnerability views of poverty; discussing measurement of extreme poverty; and highlighting feminization of formal agricultural finance. The chapter concludes with recommendations. The methodology is based on systematic reviews of relevant literature. The findings include how agriculture can play the roles expected of it and effectively empower those who are engaged in it. The chapter shares the view that majority of those engaged in agriculture in most developing countries are women, and that poverty has a feminine face and so advances the feminization of formal agricultural finance interventions. The chapter is concluded with relevant recommendations.

DOI: 10.4018/978-1-5225-3059-6.ch001

INTRODUCTION

Agriculture provides employment and income for over 90% of the Nigerian population (FMARD, 2016). It also provides food and raw materials for the people and industries. Agriculture provides foreign exchange earnings as the products are exported either as primary and or secondary goods to overseas countries. By playing the role of provision of employment, food, raw materials for industries, and foreign exchange, a well-developed agriculture holds the possibility of reducing poverty. It should be pointed out that the foregoing are the traditional roles of agriculture; in the Nigeria context as in many more developing countries, agriculture plays these roles in vary degrees. In this chapter the objective is to operationally define agriculture, conceptualize the terms "development" and "economic development", ex-ray the expected roles of agriculture in the development process, examine the performance of the agriculture sector, examine the role of agricultural credit in agricultural development, examine the concept and measurement of extreme poverty.

In this section, we provide broad definitions and discussions on the role of agriculture in the development process and subsequently incorporate views of others by way of literature review.

AGRICULTURE DEFINED

Agriculture is the science, art and occupation of conservation and cultivation of plants, animals and fungi, and other life forms for the ultimate use of humans. According to Chantrell (2002), agriculture is the cultivation and breeding of animals, plants, fibre, bio-fuel, medicinal plants and other products used to sustain and enhance human life. The definition of the term agriculture also includes the financing, processing, marketing, and distribution of agricultural products. Agriculture is a science to the extent that it involves the empirical application of the knowledge and facts about the structure and behavior of the natural and physical world. It is an art in so far as it benefits from the ability or skill developed through training and practice. It is an occupation because it is a source of livelihood to many who are engaged in it. For a developing a country such as Nigeria, the practice of agriculture is strategic to the development process. Let us briefly elaborate on the concept of development to better appreciate the role of agriculture in an economy.

THE CONCEPTS OF DEVELOPMENT AND ECONOMIC DEVELOPMENT

Development is a process of improving the conditions in which human beings live (Ibeanu, 2007). According to Rodney (1982), development as a process first involves improving man's understanding of nature(science); second applying this understanding to creation of tools and other implements that improve man's working conditions and living environment (technology); and third, equitable organization of work and rewards (social relations of production).

The more commonly occurring term among development practitioners, economic development, refers to the process whereby a nation's real per capita gross national product, GNP, or income increases over a sustained of time through continuing increases in per capita productivity (Salvatore and Dowling, 1977). However, per capita GNP data neither sheds any light on the income distribution within the nation nor on the level of general well being.

In an attempt to define economic development by first defining economic growth, Alimi (2012) conceptualized economic growth as an increase in the value of overall economic activities (measured by GNP) of a nation. According to Alimi (2012), economic development is economic growth plus positive change in socio-economic characteristics of health, education, living conditions, security and expanded range of opportunities for the citizenry. It is argued that the prerequisite for economic development is economic growth to generate the purchasing power for economic development. For economic growth and development to occur, there must be improvement in agriculture which is the main income source for majority of people in such economies as that of Nigeria; that is, there must be agricultural development.

THE EXPECTED ROLES OF AGRICULTURE IN THE DEVELOPMENT PROCESS

The Federal Government of Nigeria (FGN 1988) in its agricultural policy outlined the roles of the agricultural sector to include the following: providing adequate food for an increasing population; supplying adequate raw materials to a growing industrial sector; constituting the major source of employment; constituting the major source of foreign exchange earnings and providing a market for the products of the industrial sector. These roles are compatible and consistent with the Federal Ministry of Agriculture and Rural Development

(FMARD) (2016), which stated the vision for agriculture is to work with key stakeholders to build an agribusiness economy capable of delivering sustained prosperity by meeting domestic food security goals, generating exports, and supporting sustainable income and job growth. Let us therefore take a closer look at the expected role of agriculture in the development process.

Supply of Food

Agriculture is the only source of food for human beings. Hence it is expected to supply food for the people and with it aid in the attainment of food security and self-sufficiency in food production. Agricultural productivity can be construed of as the index of the ratio of the value of total farm output to the value of the total inputs employed in the farm production. A related term, production efficiency, means the attainment of production goals without waste. Different components of efficiency are recognized in development economics literature namely allocative efficiency, and economic efficiency. Technically efficient production process is attained if a firm produces as many output as possible from a given set of inputs or if it uses the smallest possible amount of inputs for a given level of output and input mix. Allocative efficiency shows the ability of a firm to use inputs in optimal proportions, given their respective market prices. The product of allocative efficiency and technical efficiency, referred to as economic efficiency, is defined as the ability of a firm to produce a specified amount of output at minimum cost. For agriculture to play its role as a source of food supply, agricultural productivity must be efficient.

In terms of consumption, food consumption enables the formation of human capital and the resultant better quality of labour force enhanced economic productivity. Availability and affordability of food is a necessary condition for peace. In a situation of rising food prices and attendant hunger and violent protests, food supply is the cause as well as the remedy for the conflict. Neglect of agriculture in the development process can lead to insufficient supply of food and with it food crisis—a desperate situation in which the demand for food is growing faster than the population. Gap created by the difference between demand and supply of food in developing countries is usually filled with food importation with the consequent drain on the foreign exchange reserves.

In the views of Eboh (2011), food is both a consumption and investment good. As a consumption good, it is a biological necessity. As an investment good, food consumption engenders the formation of human capital for improved labour force and improved economic activity.

Adequate food ensures peace, law, and order. This is true to the extent that rising food prices and hunger have precipitated angry and violent protests and in extreme cases regime change. For instance, according to Kron (2011), on Thursday, 21 April, 2011, riots broke out in downtown Kampala as another round of street demonstrations over commodity prices. Kron (2011), quoted Dr Green, a Ugandan expert in the London School of Economics as stating that Africa has a long history of urban food riots that brought down governments and made reference to Burkina Faso where a riot by soldiers was sparked by food riot. According to Gros, Gard-Murray and Bar-Yam (2012, prior to 2008, violence in Yemen can be attributed to inter-group conflict between ethnically and religiously distinct groups; starting in 2008, increasing global food prices triggered a new wave of violence that spread to the endemically poor southern region with demands for government change and economic concessions. Gros, Gard-Murray and Bar-Yam (2012), further posited that in Yemen, the food riots in 2007-08 and 2010-11 were not a local phenomenon, but affected a broad spectrum of regions in Africa(Sudan, Tunisia, Somalia, Cote d'Ivoire, Egypt, Mozambique, Cameroon, Mauritania, Uganda, Malawi, Kenya etc) and the Middle East (Syria, Iraq, Yemen, Oman etc) as well as Haiti and India and argued that social unrest is induced by the government's perceived failure to provide food security to the population.

It should also be pointed out that peace, law and order ensure supply of food. Agriculture is severely affected in times of political and economic instability. It is argued that in times of conflicts and political instability, farmers are not only conscripted into armies or militias but also face the added danger of their farmlands being turned to warfronts thereby depriving agriculture of workforce, consequent abandonment of the farm and farmlands and so negatively affecting food production.

Source of Raw Materials

In addition to being the source of food, agriculture provides raw materials for the manufacturing industries. If enough attention is not paid to agriculture and supplies of raw materials are not forthcoming, industries suffer. The alternative is the importation of the requisite raw materials, which like food importation drains the country's foreign earnings. Stable supply of industrial raw materials tends to be guaranteed where a nation is able to obtain the supplies from locally produced agricultural products. Furthermore, foreign exchange is conserved and at the same time dependence of local industries on imported inputs is reduced.

Foreign Exchange Earning Role of Agriculture

Agriculture is expected to be a major foreign exchange earner. Agricultural exports rake in foreign exchange with favorable effects on the balance of payment. Because primary products are the major source of export earnings in most developing countries, agriculture not only provides this resource but also provide foreign exchange earnings as a result.

Source of Employment and Income

Agriculture is the largest sector in most developing countries. In Nigeria, for instance, the agricultural sector is the largest employer. Agriculture contributes 42 percent of Nigeria's GDP, provides employment and a means of livelihood for more than 60 percent of the productive labour force. However, agriculture receives less than 10 percent of the annual budgetary allocations.

It should be stated that agriculture is expected to play these roles in the development process, the crisis of food production, and food insecurity in Nigeria are ever present. It explains the persistence of poverty. The loss of food sovereignty and the dependence on food importation is also making the country quite susceptible to fluctuations in global food crisis. Underfunding of agriculture is central to the crisis (お菓子の食べ過ぎの影響|食べ過ぎて太る食事の量はどのくらい?., n.d.).

PERFORMANCE OF THE AGRICULTURE SECTOR

This section focuses on the performance of the agriculture sector in the Nigerian economy. It is discussed under the different headings of the role of agriculture in the economy.

Supply of Food

The sector has not been sufficiently able to deal with the problem of food supply. The less than satisfactory performance of the agriculture sector in the supply of food raises the problem associated with food security. By definition, food security is attained if an individual has access to adequate quantity and quality of food for healthy and active life. Anon (2009) quoting Food and Agricultural Organization (FAO) defined food security as a condition when all people, at all times, have physical and economic access to sufficient, safe and nutritious food to meet their dietary needs and food preferences for an active health life.

Central for the existence of food security are *availability* of staple food supplies, *affordability* (access by all) and *adequac*y (in terms of quality and nutritional value). Estimates put the number of hungry people in Nigeria at over 53 million, which is about 30 percent of the country's total population of roughly 170 million. Nigeria was self-sufficient in food production and also a net exporter of food to other parts of the world in the 1950s and 1960s.However, the discovery of crude oil and the huge revenue from the petroleum sector encouraged the neglect of the agricultural sector and made Nigeria a net importer of food. By 2009 for example the federal ministry of agriculture estimated that Nigeria was spending over $3billion annually on food imports. Agriculture, which contributed 40.1 per cent to GDP in 2013, is still in a sorry state: Nigeria imports its major foods like rice, fish, poultry, and even palm oil. It is argued that underfunding in this regard is central to the crisis of food production, and food security in Nigeria. According to Federal Ministry of Agriculture and Rural Development, (FMARD) (2016) Nigeria still imports about $3 to $5 billion worth of food annually, especially wheat, rice, fish and sundry items, including fresh fruits. Specifically, rice imports still exceed $1 billion per annum. Growth in food production remains limited due to gaps in input supplies.

For agriculture to effectively play the role of food supply, it is estimated that agricultural production must increase by at least 4-6% per annum on a sustained basis to meet the food requirements of a country whose population is growing at a rate of 3% per annum. According to Eboh (2011), agriculture has witnessed relatively improved, but the growth has not translated into significant gains in food and nutritional security. The Global Hunger Index 2008-9 ranked developing countries on a scale of 0-100 with 0 indicating the absence of hunger in a specified country. Nigeria's 2008 ranking was in the 10-19 range and labeled serious. It is estimated that the protein intake is about 18gms per capita per day as opposed to the recommended level of 35gms per capita per day. According to FMARD (2010), domestic production of beef is estimated at about 672,000MT far less than the estimated annual demand which stands at 1,008,000 MT.

Supply of Raw Materials

The supply of agricultural raw materials from local sources to the industries that need them has been sufficiently deficient. According to Okafor (2015), Nigeria is known for its wealth in raw materials and solid minerals, but only 46.71% of raw materials are sourced within the country. Okafor (2015) further stated that local manufacturers outsource raw materials from other African countries, an arrangement which directly affects workers in the intermediate segments of the local economy such as farmers and workers in the solid minerals sector and animal skin firms. Okafor (2015) concluded that probable poor quality of raw materials and their insufficient supply within the country account for this arrangement. FMARD (2016) categorically stated that Nigeria has not exploited global markets for its key foods e.g. cassava for a variety of reasons including quality, market knowledge and financing issues. It should be stated that Nestle Nigeria Plc has a record of buying raw materials for its seasoning, cereal and beverage product from local farmers. For instance, in 2013 Nestle bought over 21,450 tonnes of maize, sorghum and soybean from farmers under its Quality Grains Improvement Project. A breakdown of the figure showed that in 2013, the company purchased 9,780 tonnes of maize, 5,340 tonnes of soybean and 6,330 tonnes of sorghum from farmers. In order to ensure a sustainable supply of safe, high quality grains to Nestle's factories and improve the livelihood of smallholder farmers in Nigeria, the company launched Grains Quality Improvement Project in 2009.

The problem of supply of raw materials to industries is somewhat related to that food. According to Vaughan, Afolami, Oyekale, and Ayegbokiki (2014), Nigeria is a net importer of such core food commodities as rice, sugar, wheat flour, fish, milk, etc. Food supply is inadequate, raw materials requirements for industries are also not fully met; as a result, some industries rely on foreign as opposed to local sourcing of raw materials. While deliberate and workable policies aimed at mitigating importation dependency are called for, small-scale farmers should aim at producing the right quantity and quality of goods and services for the market.

Foreign Exchange Earning Role of Agriculture

According to Shaib, Aliyu and Bakshi (1997), between 1960 and 1970, agriculture successfully played the role as a foreign exchange earner. Within this period, agricultural exports on the average accounted for almost 62% of the total value of exports. Agricultural exports took a detrimental and dismal downturn in the 1971-1980 periods. During the 1971-1980 era, the export of petroleum came to prominence. Export earnings from the petroleum sector effectively dwarfed the earnings from agriculture which stood at less than 10% of total export earnings.

In terms of specific crops earnings, Nigeria exported 167,000 metric tons of palm oil in 1961. This export earning took a huge tumble in 1970 to a dismal 8.000 metric tons in 1970. There were no palm oil exports in 1980 and 1990. Nigeria's palm oil exports earning recorded 8,000 metric tons in 2000; in 2008 it was 25,000 metric tons.

Nigeria's share of world's shelled groundnut exports in 1961was 42% totaling 502,000 metric tons. This plummeted to 291,000 metric tons in 1970. There was nil export of shelled groundnut in the 1980-2008 period.

According to Nwodo (2017), revenue in the world today is promoted by two main sources namely, human capital development leveraging on technology to drive the critical sectors of the economy and agriculture. Explaining further Nwodo (2017), stated that ten years ago the top ten companies in the world were the likes of Exxon Mobil, Shell, and Total. Today the top eight companies in the world are represented by technology related companies. They include Apple, Microsoft, Google, Facebook, and Amazon.

Nwodo (2017), further observed that The Netherlands is the 18th largest economy in the world and has a land area of about 33,900 square kilometers. Niger State, one of the states in Nigeria, has about 74,000 square kilometers. The Netherlands annually derives over $100 billion from agricultural exports, contributed mainly by vegetables and dairy. Nigeria's oil revenue has never in any one year reached $100 billion. This is a pointer to the huge unexploited revenue potentials in Nigeria's agricultural sector.

Export-oriented agricultural sector should be the goal of Nigeria. For one thing, an export-oriented agricultural economy diversifies the economy and increases foreign exchange reserves. For another, an export-oriented agricultural sector has a stabilizing effect on the macro-economy.

Source of Employment and Income

Nigeria's unemployment rate tends to be moving upwards over the past ten years. The youth are mostly affected by high incidence of unemployment as the unemployment rate is over 50%. It is estimated that 4 million young people enter the workforce every year with only a small fraction able to find formal employment. In terms of yearly averages, the unemployment rates for 2006, 2007, 2008, 2009 and 2010 are 12.3%, 12.7%, 14.9%, 19.7%, and 20.6% respectively. The unemployment rate in Nigeria increased to 14.2 percent in the last quarter of 2016 from 10.4 percent in 2015 (Nigeria Unemployment Rate 2006-2017 I Data I Chart I Calendar I Forecast., n.d.). The unemployment rate was higher for persons between 15-24 years old (25.2 percent), women (16.3 percent) and in rural areas (25.8 percent) Unemployment Rate in Nigeria averaged 9.76 percent from 2006 until 2016, reaching an all-time high of 19.70 percent in the fourth quarter of 2009 and a record low of 5.10 percent in the fourth quarter of 2010.The unemployment rates in the geopolitical zones in Nigeria range from 9.0% in the North Central to 29.0% in the North Central. The unemployment rates for North East, South and South West zones are 20.0%, 18.0% and 15% respectively.

It should be noted that the North Central Zone is the indisputable food basket of the nation. Interestingly, this zone has the least rate of unemployment. This is indicative of agriculture's ability to solve challenge of unemployment in the country as can be seen by the fact that the breadbasket region i.e. North Central Zone, has the lowest unemployment rate in the country.

It should be pointed out that agriculture as practiced in Nigeria is still dominated by smallholder farmers. These smallholder farmers still rely on traditional farming practices and to that extent unattractive to young people. As a result, agriculture in Nigeria has not yet met the expectation of employment generation to the labour force.

THE ROLE OF AGRICULTURAL CREDIT IN AGRICULTURAL DEVELOPMENT

Agricultural credit can be described as the process of obtaining control over the use of money, goods and services in the present in exchange for a promise to repay an agreed amount at a future date. Agricultural credit transactions can be in form of loans, promissory notes, bills of exchange and bankers' acceptances. These forms of financing are adapted to the specific financial needs of farmers; the needs of the farmers in turn determined by planting, harvesting and marketing cycles and the scale of the farming business.

Agricultural finance is critical for producers of all sizes (from smallholder farmers, medium size farmer and larger commercial farms) as well as to properly-functioning input supply markets, processors and traders (FMARD, 2016). Finance fixes four functions for a commercial economy: provision of efficient means of exchange, encourages more efficient resource allocation through transferring claims on resources from surplus units to deficit units, provides for the transformation and redistribution of risk among units, serves as a tool in economic stabilization activities. These four functions of finance to a large extent define the role of agricultural credit in the agricultural development process.

The role of agricultural credit in the development process can be graphically explained by employing the vicious circle argument. It has been argued that low income leads to low savings, low savings give rise to low investment, low investment translates to low productivity, low productivity engenders low income and so the cycle continues. This vicious circle of poverty can be broken by the external injection of production capital which will expectedly and ultimately increase productivity, increase income, increase savings and investment and lift the entity concerned out of the poverty level. In agricultural production process, this external injection of capital comes in the form of agricultural credit. It therefore stands to reason that agricultural credit, when properly administered, plays a role in breaking the vicious circle of poverty.

As the search continues for the most effective and sustainable approaches to end extreme poverty, it is germane to periodically reassess our common understanding of the concept of poverty. This periodic reassessment is borne out of the fact that the concept of poverty is not only multidimensional but also dynamic and in a constant state of flux.

Related to the multidisciplinary conceptualization of poverty, World Bank (2000) presented the now traditional view of poverty as encompassing not only material deprivation measured by an appropriate concept of income and consumption but also low achievements in education and health. In the views of Chambers (2006), the concept of poverty has at least five clusters of meanings. The first is income-poverty or, its common proxy, consumption-poverty. The second is material lack or want. Besides income, material lack or want includes lack of or little wealth, no or poor access to services etc. A third cluster of meanings is conceptualized as capability deprivation. A fourth cluster generally focuses on the multidimensional view of deprivation. According to Zupi (2007), deprivation refers to a lack of well-being not only in terms of material goods but also as related to emotional and psychological aspects of life. The fifth cluster takes into account the analysis and views of the object of the definition and description- the poor. The fifth cluster is the multiplicity of the descriptions of the poor themselves. The fluidity of the concept of poverty is obvious when it is realized that in 1990, 43% of the population of developing countries lived in extreme poverty then defined as subsisting on $1 a day. By 2010 it was 21% or 1.2 billion; the poverty line was then $1.25. The dynamic nature of poverty is presented by Zupi (2007) who stated that poverty becomes chronic over time; the exact length of time being somewhat arbitrary.

The now traditional view of poverty based on income measures. However, there are alternative measures and conceptualizations of poverty other than income poverty. One of such conceptualizations is the human person vulnerability viewpoint.

THE CONCEPT OF EXTREME POVERTY

Extreme poverty is defined as a condition characterized by severe deprivation of basic human needs, including food, safe drinking water, sanitation facilities, health, shelter, education and information (Extreme poverty, n.d.). It depends not only on income but also on access to services. It is an economic condition of lack of both money and such basic necessities for successful life as food, water, education and shelter.

In many developed countries, the official definition of extreme poverty used for statistical purposes is based on relative income. For instance, in the post-2015 development agenda, the overall goal of eradicating extreme poverty by 2030 means reducing the number of people earning less than $1.25 a day by 50 million each year. These estimates of income poverty are essentially based on household income and expenditure surveys. The next section examines the strengths and weaknesses of income measure of poverty.

SOME STRENGTH AND WEAKNESSES OF INCOMETRICS

To the extent that income poverty estimates are based on nationally representative samples they allow inferences about the conditions and evolution of poverty at the national level (WB, 2000). On the other hand, survey design varies between countries and over time, often making comparisons difficult. It is also argued that income based poverty statistics measure inequality rather than vulnerability or hardship. There is the need to interpret poverty from such perspectives as basic needs, participatory assessment, human rights, social exclusion, capabilities and vulnerabilities.

VULNERABILITY VIEWS OF POVERTY

Vulnerability is the risk that a household or individual will experience an episode of income or health poverty over time (WB, 2000). It also means the probability of being exposed to a number of other risks such as violence crime, natural disasters, and being pulled out of school. According to Coudouel and Hentschel (2000), vulnerability encompasses not only income vulnerability but also such risks as those related to health, those resulting from violence, and those resulting from social exclusion – all of which can have dramatic effects on households.

Vulnerability refers to the relationship between poverty, risk, and efforts to manage risk (Alwang, Siegel, & Jorgensen, 2001). Alwang, Siegel and Jorgensen, (2001) decomposed vulnerability into several components of a risk chain: a) the risk, or risky events, b) the options for managing risk, or the risk responses, and c) the outcome in terms of welfare loss. Focusing on household vulnerability Alwang, Siegel and Jorgensen (2001), argued that households are vulnerable to suffering an undesirable outcome, and this vulnerability comes from exposure to risk and posited that social actions can reduce risk or exposure to risk.

Vulnerability can also be defined as the probability of an acute decline in access to food, or consumption, often in reference to some critical value that defines minimal values of human well-being (WFP 2002). Vulnerability influences a household's behavior and means of coping with its current wealth and livelihood circumstances; it also has an impact on the state of food security at the individual and household levels. Environmental and situational factors, such as regional climactic patterns or national economic trends, bring additional levels of influence that interact with the circumstances of individual households to further affect food security

Poverty reduction and/or eradication strategies to all intents and purposes are interlinked with strategies aimed at reducing socioeconomic risks and vulnerabilities. Poor rural households that mostly rely on agriculture for their livelihoods are often affected by limited access to resources, low agricultural productivity, poorly functioning markets and repeated exposure to risks.

DEFINITION AND MEASUREMENT OF EXTREME POVERTY

The attempt to define poverty in non-income metric terms should essentially be guided by issues of global concern that are sufficiently reflected in local livelihood aspirations. It is against this backdrop that extreme poverty can be defined as the susceptibility and vulnerability of an individual or a household to the adverse effects of extreme weather conditions on account of lack of effective mitigation and adaptation measures to climate change.

In Nigeria for example, the unmitigated disaster that is the environment of the Niger Delta due to oil exploitation is replicated in the abandoned tin mines of the plateau and the coal mines of Enugu, Kogi and others. Desertification threatens at least 30% of the States of Nigeria. Gully erosion affects States in the South East, South and the North Central, while coastal erosion affects the entire national coastline.

The protection of our biodiversity is a major challenge going by the continued habitat loss, pollution of water bodies and heavy deforestation. Remaining forests in Cross River, Edo, Ondo, Ekiti and other States must be protected if the country is to continue to have primary forests. Loss of grasslands is equally presents environmental challenge with far-reaching implications.

Agriculture remains the main source of livelihoods for most rural communities in the rainforest region of Nigeria. Environmental degradation as a result of climate change is expected to have greater negative impacts on poorer farm households as they have the lowest capacity to adapt to changes in climatic conditions and more vulnerable to hazards of environmental degradation. Adaptation measures are important to help households to better face extreme weather conditions and attendant variations (Adger *et al.*, 2003). Adaptation measures have the potential to contribute to reductions in negative impacts from environmental degradation, changes in climatic conditions as well as other changing socioeconomic conditions, such as volatile short-term changes in local and international markets (Kandlinkar & Risbey, 2000).

Measuring vulnerability is especially difficult. It is not impossible. According to WB (2000), household panel data can provide basic information with which to capture and quantify the volatility, susceptibility and vulnerability that households say is important. However, efforts should be made to be proactive in determining vulnerability as opposed to vulnerability only after the fact. Furthermore, it is neither feasible nor desirable to capture vulnerability in a single indicator.

Evidently, the aim of agricultural development is to encourage farmers to make substantial investment in agriculture and stimulate increase in productivity. Because the resource endowments of majority of small-scale farmers cannot economically sustain capital formation and enterprise expansion, the capital requirement for investment in agriculture must be injected from outside the resource and financial capabilities of the farmers.

One of the factors attributed to the declining productivity of the agricultural sector is farmers' limited access to credit facilities. Acquisition and utilization of credit for agricultural purposes promote productivity and consequently improved food security status of a community. Agricultural credit, when properly managed, not only promotes productivity and ultimately increasing the farm income of farmers, it also plays a role in enhancing the standard of living of the farmers as a result. This means that there is increase in the level of wealth, comfort, material goods and necessities available to the farmers whose productivity increased on account of utilisation of agricultural credit.

Agricultural credit is not only important for improving farmers' productivity; it also aids in accelerating agricultural development. Increase productivity depends on adoption and technical efficiency of improved farming technologies. Agricultural credit enables farmers to have access to production inputs, plant alternative crops and adopt modern farm technologies. Agricultural credit therefore plays important role in agricultural development and is a key to agricultural modernization. Thus, an agricultural credit service is considered an important component of the agricultural development programme.

Related to the role of agricultural credit in accelerating agricultural development is that of enabling farmers to have access to production packages and try them out as recommended. It is rightly argued that agricultural credit has been discovered to be a major constraint on the intensification of both large and small-scale farming. Agricultural inventions, innovations, technologies and packages which are products of agricultural research and development would have no utility to the society, if they are not put into actual practice on the farm. Adoption of such technologies involves a lot of capital outlays, which the farmers ordinarily cannot finance from their savings. In Nigeria, farmers require an appreciable amount of capital for efficient farm operations. Capital is also needed for the acquisition of such agricultural purchased inputs as fertilizers, farm tools, implements, pesticides, insecticides and herbicides. Agricultural credit therefore enables farmers to access and adopt improved technologies.

It should be pointed out that credit is not only needed for farming purposes, but also for family consumption expenses especially during the off-season period. Nigerian agricultural economy requires credit for the purchase or rent of farm lands, farm machines, building of farm structures irrigation, storage facilities, hiring of labour. Credit is needed for farm improvement, marketing of farm products, veterinary services, and recurrent. Agricultural credit has the potentials to enhance the efficient resource allocation, reduce post-harvest waste and stabilize farm prices.

The role of agricultural credit in agricultural development should be discussed with all caution. For one thing, credit is neither a magic wand nor a panacea for the drawbacks of agricultural production. Agricultural credit loses its essence if it does not the equity and efficiency objective as expressed in enterprise expansion. The enviable role of agricultural credit notwithstanding, serious loan recovery problems persist. For instance, some agricultural credit institutions have failed as a result of lack of repayment of loans. Other problems preventing agricultural credit from accomplishing its expected roles include the reluctance of institutional financial intermediaries to lend to small-scale farmers; political affiliations continue to dictate agricultural lending procedures and beneficiaries and most of the government owned lending agencies flounder on account of their costs of lending and defaults exceeding returns. For agricultural credit to fill the bill in the agricultural development process, some popular misconceptions about what agricultural credit really is should be dropped.

Too many people think of agricultural credit as a productive input instead of a general claim on any good or service in the market. It is therefore the duty of the discerning loan beneficiary to channel the agricultural loan that is obtained to economic, productive and remunerative activities for achieving efficiency and effectiveness.

Additionally, because like financial instruments, agricultural credits are fungible and divisible; so the notion of control that many beneficiaries have over them can be described as illusory. An instrument is fungible if it is of such a nature or kind as to be freely exchangeable or replaceable in whole or in part for another of like nature or kind. The financial instruments are fungible to the extent that once in the hand of the beneficiary, it can be put to any use other that it is originally meant. For instance, a loan amount in the hand of a beneficiary for agricultural production might end up being used to settle hospital expenses. In a related sense, divisibility characteristics is evidenced by an agricultural loan amount ending up being thinly spread for other uses than originally intended be they agricultural or non-agricultural, productive or consumptive. Other examples include seed and grains for planting ending up on the dining table instead of being planted in the field; purebred piglet given to farmers on the condition that they return a piglet from the first litter and the piglet being served for dinner before any piglet is produced.

From the foregoing discussions, it is clear that the role of agricultural credit can only be realized if the borrower, the lender and relevant stakeholders clearly understand the nature and meaning of the financial instrument, good or service. It is only then that we can realize the role of agricultural credit in agricultural development.

FEMINIZATION OF FORMAL AGRICULTURAL FINANCE

This section focuses on formal agricultural credit delivery targeted at the women farmers. This is borne out the fact that the role women play and their position in meeting the challenges of agricultural production and development are important and visible. Increases in agricultural productivity are central to growth, income distribution, improved food security and alleviation of poverty in rural Africa (FAO, 2002) including Nigeria. Women make up some 60-80 percent of agricultural labour force in Nigeria (World Bank, 2003) and produce two-thirds of the food crops. Women account for 75 percent of the farming population in Nigeria, working as farm managers, and suppliers of labour. Though women constitute a large portion of the farming population, women's possibilities in agriculture are hindered by formal and traditional rules. Generally, the extent of gender involvement in agricultural production varies across ethnic groups in Nigeria (SAHEL, 2014). Nigerian women contribute close to 70 per cent of agricultural workforce yet get less of accruing returns (Premium Times, 2015).

However, female farmers in the country are among the voiceless, especially with respect to influencing agricultural policies. Such policies, which are aimed at increasing food security and food production, tend to either underestimate and totally ignore women's role in both production and the general decision-making process within the household (Ogunlela & Mukhtar, 2009).

According to SAHEL (2014), the key constraints affecting women's ability to improve efficiency in agriculture include: Access to financing: **i**n Nigeria, women farmers receive less than 10% of the credit offered to small-scale farmers; access to information and training; access to inputs: this is on account of high cost farm inputs in the open market; access to land: women in Nigeria generally own less land than men due to traditional authority. Discussions on the concept of feminization of formal agricultural finance may not be complete without reference being made to the original concept of feminization of poverty.

Diana Pearce coined the phrase "feminization of poverty" in 1978. Feminization of poverty refers to the tendency for female proportion of the population with lower incomes than men. It is the increasing female proportion of the population on low incomes or in poverty (sociologyindex.com/feminization_of_poverty.htm). It is a social process whereby the incidence of poverty among women is higher than among men. Chant (2006), viewed it as a phenomenon that women represent disproportionate percentages of the world's poor. According to Chen et al. (2005), it is burden of poverty borne by women especially in developing economies. It simply means that an increasing proportion of the world's poor are female. Majority of the 1.5 billion people living on 1 dollar a day or less are women. Moghadam (2005), argued that although the claim that the majority of the world's poor are women cannot be substantiated, the disadvantaged position of women is incontestable. The disadvantaged position of women tends to give support to the sex-linked characteristics of poverty. According to United Nations Department of Public Information (UNDPI) (2000), feminization of poverty is a phenomenon of the widening gap between women and men caught in the cycle of poverty.

According to Moghadam (2005), feminization of poverty was coined to describe the expansion of poor or low-income female-headed households in the United States (and Latin America); in the developing world and the transition economies and is associated with the social costs of structural adjustment and market reforms. In answer to the questions "Is the expansion of female-headed households behind the growth of women's poverty?" and "Are female-headed households the poorest of the poor across the world?", Moghadam (2005), stated that the available evidence is inconclusive for most regions, but strongest in the United States, where the feminization of poverty is associated with the increase of female-headed households among low-income women in terms of inadequate social supports. Moghadam (2005), further observed variation across regions in the economic status of female-headed households, based partly on the social policy or political regime, and partly on women's access to employment and property. Women living in poverty are often denied access to critical resources such as credit, land and inheritance.

Moghadam (2005), observed that women's greater vulnerability to poverty is associated with the existence of gender inequalities in household resource allocations and decision-making, in public policies, and in legal codes. According to Moghadam (2005), women are more vulnerable to poverty due to lack of property rights in land or access to employment, lower wages, illiteracy, early marriage and childbearing, lack of rights in divorce, incomplete education, and lower wages.

Moghadam (2005), concluded that globally, women are especially severe victims of poverty in at least three ways. First, gender inequalities put women at a distinct disadvantage vis-à-vis men and in the face of a range of impoverishing and exploitative conditions. Second, women work longer hours than men at both productive and reproductive activities, and still earn less. Third, the capacity of women to lift themselves out of poverty is circumscribed by cultural, legal, and labor-market constraints on their social and occupational mobility. As a result, poverty tends be more chronic among women and more transient among men.

UNDPI (2000), recommended that empowering women is an essential factor in releasing billions of people caught in the web of poverty and deprivation adding that the provision of credit particularly microcredit is a sure and successful strategy for poverty eradication. Moghadam (2005), recommended that reducing the deprivation experienced by poor women and the greater difficulties women face in lifting themselves, and their children, out of poverty across the globe. The recommendations included well-designed social programs such as distribution of land, training programmes, job-creation, employment services, access to housing, childcare facilities, and reform of discriminatory laws. Moghadam (2005), argued that social welfare mechanisms need to be strengthened for all, but an effective long-term anti-poverty strategy should address itself to the enhancement of women's entitlements and capabilities, whether in male-headed or female-headed households.

Agricultural credit is a necessity in increasing the capital base of women famers. Women farmers' lack of credit facilities is the most pressing problem women face in agriculture. According to Adebyo and Tunde (2013), majority of women farmers cultivate small plots because they do not have access to credit with which to purchase larger plots and farm inputs for increased production. According to Sanusi (2012), the key barriers to the growth of women-owned enterprises in developing countries include lack of collateral, absence of credit histories, biased attitude of banks towards women, lack of property rights and control over assets.

In the views of Buvinic and Gupta (2013), female headed households in most rural areas of developing countries are prone food insecurity. Food insecurity in the households has been attributed to various forms of discrimination, to poor access to productive inputs, assets and services (Amaza et al. 2006; World Bank 2013).

In order for agriculture to advance in Nigeria, gender specific policies and services tailored to women in the value chains should be developed (SAHEL, 2014). To realize growth of agribusiness entrepreneurs, majority of who are women, Premium Times (2015), recommended five interventions: training, increasing access to inputs, financing, enhancing market links and expanding co-operative programmes like women co-operatives.

In the views of the Nigeria Country Director, African Development Bank, According to Premium Times (2015), the bank's 10 years strategy pan (2012-2022) has placed high emphasis on gender equality and mainstreaming as prerequisites for African economic transformation. The AfDB and partners plan to sustainably and proactively support African women by developing unique tailor-made projects for women-led and women-dominated businesses in agriculture.

SOLUTIONS AND RECOMMENDATIONS

For agriculture to play its expected role in the development process, the government should provide the enabling environment through incentives, subsidies and efficient credit facilities. Based on the concept of feminization of poverty, the government should consciously embark on what may be referred to as feminization of supervised credit facilities to farmers. Given the disadvantaged and deprived position of women including women farmers, special empowerment schemes for them will go a long way in aiding improvement in their farm incomes. Furthermore, agriculture supports a greater percentage of the people; it is obviously a private sector activity. So, private sector investments in agriculture are called for to enable agriculture to play the expected roles to the people and to the economy of Nigeria.

FUTURE RESEARCH DIRECTIONS

Majority of the population of Nigerians are engaged in agriculture, yet agriculture cannot adequately feed the people to the extent that a lot of money is annually committed to the importation of even the food items we produce. The scenario has been blamed on increases in the population not being matched with increases in food production, neglect of agricultural sector etc. The issue is that this problem has faced Nigeria for decades without a clear-cut solution. It is therefore recommended that further research be carried out on how the money annually spent by Nigeria on the importation of food can effectively directed to financing the local production of such imported food items.

CONCLUSION

For a developing country such as Nigeria, the practice of agriculture is strategic to the development process in general and in improving the conditions in which human beings live. The roles of the agricultural sector include providing adequate food for an increasing population; supplying adequate raw materials to a growing industrial sector; constituting the major source of employment; constituting the major source of foreign exchange earnings and providing a market for the products of the industrial sector. Agriculture as practiced in Nigeria is still dominated by smallholder farmers who still rely on traditional farming practices and to that extent unattractive to young people. Agriculture in Nigeria has not yet met the roles expected of it including employment generation to the labour force.

REFERENCES

Adebayo, A., & Tunde, A. M. (2013). Challenges of women in urban agriculture in Kwara State, Nigeria. *Sustainable Agriculture Research*, *2*(3).

Adger, Huq, Brown, Conway, & Hulme. (2003). *Adaptation to Climate Change in the Developing World.* Progress in Development Studies No. 3.

Alimi, T. (2012). Small or Large Scale Agriculture for Nigeria: Issues, Challenges and Prospects. Inaugural Lecture Series 249. Obafemi Awolowo University Press.

Alli, F. (2014). *Sourcing raw materials locally 'll create market for farmers — FG.* Retrieved October14, 2017 from https://www.vanguardngr.com/2014/08/sourcing-raw-materials-locally-ll-create-market-farmers-fg/

Alwang, J., Siegel, P. B., & Jorgensen, S. L. (2001). *Vulnerability: A View from Different Disciplines.* Social Protection Discussion Paper Series No 0115. World Bank.

Amaza, P. S., Umeh, J. C., Helsen, J., & Adejobi, A. O. (2006). *Determinants and Measurements of Food Insecurity. Nigeria: Empirical Policy Guide*. Paper presented at International Association of agricultural Economics Conference, Gold Coast, Australia.

Anon. (2009). Food Security: An Engine of Growth and Poverty Alleviation in Africa. *Proceedings o the Summit on Food Security in Africa.*

Buvinic, M., & Gupta, G. R. (2013). *Female-headed households and Female-maintained Families: Are they worth targeting to reduce poverty in Developing Countries. Economic Development and Well-being in North India, DEP No 40 STICARD.* London School of Economics.

Chambers, R. (2006). What is Poverty? Who Asks? Who Asks? In D. Rhrempreis (Ed.), *Poverty in Focus*. UNDP International Poverty Centre Rio de Janeiro.

Chant, S. (2006). Rethinking the 'Feminization of Poverty' in relation to aggregate gender indices. *Journal of Human Development and Capabilities*, *7*(2), 201–220. doi:10.1080/14649880600768538

Chantrell, G. (2002). *The Oxford Dictionary of Word Histories*. Oxford University Press.

Chen, M., Vanek, J., Lund, F., Heint, J., Jhabvala, K., & Bonner, C. (2005). *Progress of the World's Women 2005: Women*. Work and Poverty.

Coudouel, A., & Hentschel, J. (2000). *Poverty Data and Measurement. Preliminary Draft for A Sourcebook on Poverty Reduction Strategies. The World Bank.*

Eaton, C., & Shepherd, A. W. (2001). Contract farming partnerships for growth. *FAO Agricultural Services Bulletin, 145.*

Extreme poverty. (n.d.). Retrieved May 21, 2014, from https://en.wikipedia.org/wiki/Extreme_poverty

FAO. (2002). F.*A.O Report to "The new partnership for Africa's development": Land and water resources issues and agricultural development. 22nd Regional Conference for Africa*, Cairo, Egypt.

Federal Ministry of Agriculture and Rural Development (FMARD). (2010). *Global Agricultural and Food Security Support Programme*. Nigeria: FMARD.

Federal Ministry of Agriculture and Rural Development (FMARD). (2016). *The Agriculture promotion policy (2016 – 2020) building on the successes of the ATA, Closing key gaps policy and strategy document*. Federal Ministry of Agriculture and Rural Development.

Gros, A., Gard-Murray, A. S., & Bar-Yam, Y. (2012). *Conflict in Yemen: From ethnic fighting to food riots*. Retrieved October 16, 2017 from https://necsi.edu/research/social/Yemen

Ibeanu, O. (2007). Conceptualizing Peace. In S. G. Best (Ed.), *Introduction to Peace and Conflict Studies in West Africa: A Reader*. Spectrum Books Limited Ibadan.

Kandlinkar & Risbey. (2000). Agricultural Impacts of Climate Change: If Adaptation is the Answer, What is the Question? *Climatic Change, 45.*

Kron, J. (2011). *Protests in Uganda Over Rising Prices Grow Violent*. Retrieved October 16, 2017 from https://mobile.nytimes.com/2011/04/22/world/Africa/22/Uganda

Moghadam, V. M. (2005). *The 'Feminization Of Poverty' And Women's Human Rights*. UNESCO France.

お菓子の食べ過ぎの影響|食べ過ぎて太る食事の量はどのくらい?(n.d.). Retrieved February 4, 2014, from http://westafricainsight.org/articles/PDF/81

Nigeria Unemployment Rate 2006-2017 | Data | Chart | Calendar | Forecast. (n.d.). Retrieved January 16, 2017, from https://tradingeconomics.com/nigeria/unemployment-rate

Nwodo, J. N. (2017). *Restructuring Nigeria: Decentralisation for national cohesion*. Retrieved September 30, 2017 from https://guardian.ng/politics/restructuring-nigeria-decentralisation-for-national-cohesion/

Ogunlela, Y. I., & Mukhtar, A. A. (2009). Gender Issues in Agriculture and Rural Development in Nigeria: The Role of Women *Humanity &. The Social Science Journal*, *4*(1), 19–30.

Okafor, C. (2015). *Sourcing Raw Materials in Nigeria Can Aid Our Employment Market*. Retrieved October14, 2017 from http://venturesafrica.com/sourcing-raw-materials-in-nigeria-can-aid-our-employment-market/

Premium Times. (2015). *Nigerian women contribute 70 per cent of agricultural workforce – AfDB*. Retrieved November 13, 2017, from https://www.premiumtimesng.com/news/top-news/191777-nigerian-women-contribute-70-per-cent-of-agricultural-workforce-afdb.html

Rodney, W. (1982). *How Europe Underdeveloped Africa*. Ikenga Publishers Enugu.

SAHEL. (2014). *The role of women in Nigerian agriculture*. Retrieved November 13, 2017, from http://sahelcp.com/wp-content/uploads/2016/12/sahel-newsletter-volume-7.pdf

Salvatore, D., & Dowling, E. (1977). *Development Economics*. McGraw-Hill Book Company.

Sanusi, L. S. (2012). *Increasing Women's access to Finance: Challenges and Opportunities*. Paper presented at the 2nd African Women's Economic Summit, Lagos, Nigeria.

Shaib, B., Aliyu, A., & Bakshi, J. S. (1997). *National Agricultural Research Strategy Plan 1996-2010*. Nigeria: Department of Agricultural Sciences Federal Ministry of Agriculture and Natural Resources.

Upton, M. (1973). *Farm Management in Africa the Principles of Production and Planning*. Oxford University Press.

Vaughan, I. O., Afolami, C. A., Oyekale, T. O., & Ayegbokiki, A. O. (2014). An Analysis of Nigeria Food Imports and Bills. *International Journal of Economics, Commerce and Management*, *2*(9). Retrieved from http://ijecm.co.uk/

WB (World Bank). (2000). *World Development Report 2000/2001 Attacking Poverty*. Author.

WFP (World Food Programme). (2002). *VAM Standard Analytical Framework*. Rome: WFP.

World Bank. (2003). Nigeria: Women in agriculture. In *Sharing Experiences—Examples of Participating Approaches*. The World Bank Group. Retrieved from http:/www.worldbank.org/wbi/publications.html

World Bank. (2013). *Nutrition Research: Pursuing Sustainable Solutions.* Retrieved from http://reliefwed.int/sites/resources/fullreport_153.pdf

Zupi, M. (2007). The Multi-Dimensions of Poverty. Some Conceptual and Policy Challenges. *Development Society for International Development*, *50*(2), 31–34.

Chapter 2
Finance for Agriculture

ABSTRACT

Finance for funding agricultural activities can be fixed through three typical sources, namely formal, informal, and semi-formal channels. However, the small-scale farmers either by omission or commission tend to rely solely on informal sources of finance, for instance, personal savings, for funding agricultural activities. There is therefore the need to inform farmers on the array of sources of finance that are available for them for use boosting their farm investment and enterprise expansion. The objectives of the chapter include highlighting three important functions of an enterprise, defining the meaning of agricultural finance, describing sources of investment fund for agriculture, and discussing the rural finance institution-building programme. The methodology adopted is that of systematic and analytical review of relevant literature. Following from the review of relevant literature, this chapter argues that the reported reluctance of banks to lend to agriculture is dictated by the profit-making motives of the banks against the characteristic risky nature of agricultural enterprises. A way around the reluctance is government backed and efficiently targeted credit schemes. The chapter further argues that farmers should form cooperative societies and source loans from such cooperative societies at single digit interest rates. With respect to the cooperative societies, government should constantly and consciously discharge its responsibilities as they relate to supervision, monitoring, and evaluation of the activities of registered cooperative societies. The chapter observes that the formal sources of agricultural finance are capable of mobilizing a

DOI: 10.4018/978-1-5225-3059-6.ch002

large sum of money; however, in the process of financial intermediation, a relatively small fraction of the whole gets to agriculture. On their part, the informal sources tend to mobilize funds for agriculture at high and therefore unsustainable interest rates to farmers. Incentives and subsidies are therefore required for financing agriculture. It is specifically recommended that African governments should strive to commit themselves to the Comprehensive Africa Agriculture Development Programme's call for the allocation of at least 10% of the national budget for agriculture to achieve a target of 6% annual agricultural growth.

INTRODUCTION

Accessing funds for agricultural purposes can be affected through two main sources namely formal and informal channels. The formal channels are defined by their being under the complete regulatory control of the Central Bank of Nigeria. On the other hand, the informal sources are those channels that do not come under any regulatory control of the government or governmental institution. However, there are some channels of agricultural finance whose activities are partially regulated by the government. These are broadly categorized as semiformal sources. This chapter specifically highlights the three important functions of an enterprise, discuss the meaning of agricultural finance, describe the sources of investment fund for agriculture, describe the formal sources of agricultural finance and describe the informal sources of agricultural finance.

BROAD DEFINITIONS AND DISCUSSIONS

Three Important Functions of an Enterprise

The three important functions of an enterprise are finance, production and marketing. These functions are coordinated through efficient management of the resources available to the enterprise. The finance functions of an enterprise involve raising funds, generating returns from the funds raised and also paying some cost to the suppliers of the capital. The finance functions also encompass paying dividends to the owner or owners of the enterprise.

The process of transforming inputs into outputs is referred to as production. Output together with waste products, is the outcome of a production process. Inputs are those productive resources in a production process namely land, labour, capital and management. Agricultural marketing is concerned with all stages of production that enable the movement of products, goods and services from farms to consumers. The operations include assemblage of goods, transportation, storage, processing, grading and the financing of the activities. Marketing functions dictate production policy which is in turn based on the objectives of the enterprise. Furthermore, production is dictated by management finance function. So, in addition to raising funds, financial management is concerned with production, marketing and other functions within an enterprise. The finance, production and marketing functions in an enterprise are essentially non-sequential. They are instead undertaken simultaneously and continuously. Thus, finance, production and marketing are a thick tripartite thread that ties the thinking producer to the discerning consumer. Because this book is focused on the subject of agricultural finance, the interested reader can consult other texts for a deeper investigation of the concepts of agricultural production and agricultural marketing.

Meaning of Agricultural Finance

Agricultural finance refers to the economic and systematic study of the acquisition and use of capital in agriculture. In an economic sense, it can be defined as goods created by the process of investment that is capable of producing economic wealth especially in the agriculture sector. It focuses on the demand, supply, acquisition and use of funds for agricultural purposes.

At the national and state levels, agricultural finance focuses on agriculture contributions to and appropriations of the national and or state resources. It is also concerned with the role of banks and other financial institutions play in the financing of agriculture as a sector of the national economy.

At the farm level, agricultural finance focuses on the financial management of the farms as unit of production. Financial management can be construed as the managerial functions that relate to the planning and controlling of the financial resources of a given firm. Agricultural finance at the farm level can also be referred to as the study of the acquisition and use of capital in the farm in line with the aims and objectives of the farm.

Agricultural credit, a related term, refers to the process of obtaining control over the use of money goods and services in exchange for a promise to repay at a future date. Agricultural credit embraces loans and advances granted to borrowers to finance and service production activities in the agriculture sector.

The terms agricultural credit and agricultural loan are generally used interchangeably. However, credit refers to the capacity to borrow. On the other hand, the amount borrowed constitutes the loan. Other terms related to agricultural finance include microfinance, microcredit and microloan.

Microfinance can be described as banking for the poor and the core poor. It is characterized by smallness of loans and savings, absence or reduced emphasis on collateral, and simplicity of operations. It is concerned with providing financial services to the poor who are ordinarily not served by the orthodox financial institutions. Its programmes provide loans, savings and other financial services to the low income and poor people for use in small business including agriculture. Microfinance is the provision of very small loans that are repayable within short periods of time. Clients of microfinance are essentially low-income earners and households whose little assets may be used as collateral. To all intents and purposes, microfinance also includes financial products and services targeted at the poor. This targeting approach distinctively marks out microfinance from other financial services. Thus, microfinance conceptually encompasses such services as microcredit, micro-savings, micro-leasing, money transfer and payment services.

Microfinance adopts the credit-plus approach. It not only focuses on credit to the poor, but includes the additional activity of combining credit with other developmental functions such as community organizing, leadership training, family living, entrepreneurial skills and financial management.

Microcredit refers to credit directed at the poor; it is designed to enable the poor have access to production capital. Microcredit does not include other financial services such served by microfinance. The difference between microfinance and microcredit is the credit plus approach adopted by microfinance.

Loans including agricultural loans come in varied forms. Generally, can be secured loans or unsecured loans. Secured loans are loans that are supported by collateral from the borrower. The collateral, which is a mutually agreed item of higher value than the loan amount pledged by the borrower as a security for loan repayment. Items used as collateral depend on both the financial institution and the financial market in question. Unsecured loans neither are nor backed by collateral; they are based on a common understanding between the lender and the borrower.

Sources of Investment Fund for Agriculture

Investment is the act of producing goods that are not for immediate consumption. Investment goods are goods produced not for present consumption. Examples are capital goods, inventories and residential housing. Consumption is the act of using goods and services to satisfy wants.

Farmers' demand for investment fund is met through three main sources. These are formal, semiformal and informal sources. They are also known as institutional and non-institutional sources. Formal sources are defined as those whose operating procedures are strictly and statutorily standardized and are completely regulated by a government-owned institution. Examples are the deposit money banks also known as commercial banks, microfinance banks, the Nigeria Agricultural and Cooperative Rural Development Bank (NACRDB), Bank of Agriculture and the Imo state government-owned credit institutions such as the Imo Supervised Agricultural Credit Loans Scheme. Informal sources are typically not under the control of the government. Examples include money lenders, *isusu*, and rotating savings and credit associations. Semiformal sources tend to be partially under the control of a state or local authority, for example cooperative societies.

The sources of investment fund be they formal, semiformal or informal can be basically categorized as equity capital (own capital) or someone else's capital also called non-equity capital or debt financing. The amount of capital actually owned by the farmer as opposed to the value of capital borrowed is known as the equity capital. Equity capital does not require repayment. Equity capital also called own capital is mobilized from current owners of the farm firm or from outsiders who provide funds in exchange for a stake in the enterprise. On the other hand, debt capital is the funds obtained for an enterprise through borrowing and which must be paid back principal and interest as applicable within a stipulated time period. It is categorized into short term, medium term and long term debts. Short term debts are appropriated to current activities of an enterprise within a period of one year. Medium term debt is essentially used to finance medium term operations which require payback within a period of more than a year but less than three years. Long term debt tends to permit investment in fixed assets such as land, machines and buildings. The payback period is in the range of three years and above.

Let us, under the following headings, examine the main sources of finance available to a typical farm enterprise.

Formal Sources

Banks

Banks play prominent role as a source of fund for financing agriculture. Banks whether owned by the government or by private businessmen and women are financial institutions that are set up for the purpose of making profit. They are institutions for accepting, lending, exchanging and safe-keeping of money. Additionally, banks, particularly Central Banks, issue notes and transact other financial businesses such as raising letters of credit. In carrying out the function of accepting deposit form the pubic, banks through the process of financial intermediation extend credit facilities to small and medium enterprises in Nigeria. A process through which funds are channeled from surplus (depositors) units generally at cost (interest) to deficit (borrowers) units is referred to as financial intermediation.

Although it is often argue that banks are generally reluctant to grant (especially long term) loans to agricultural sector, that perceived reluctance does not remove the role from the banks. In Nigeria, there are five categories of banks that are involved in financing agriculture. They are the Central Bank, the Commercial Banks also known as Deposit Money Banks the Merchant Banks, Development Banks and the Specialized Banks

Central Bank of Nigeria

The major statutory regulatory objectives of the Central Bank of Nigeria are: maintenance of the external reserves, promotion of monetary stability and a strong macroeconomic environment, lender of last resort and financial adviser to the Federal Republic of Nigeria. As a lender of last resort, the Central Bank of Nigeria is visibly involved in credit allocations to the real sector including the agriculture sector. The role of the Central Bank of Nigeria in credit allocation to the agriculture sector is clearly seen in its activities in the Agricultural Credit Guarantee Scheme Fund.

The Agricultural Credit Guarantee Scheme Fund

The Agricultural Credit Guarantee Scheme Fund (ACGSF) was established by the Federal Military Government of Nigeria under the Agricultural Credit Guarantee Scheme Fund Decree 1977 (Decree No. 20) and as amended on 13thJune, 1988. According to CBN (2016), the Agricultural Credit Guarantee

Scheme Fund (ACGSF) encourages lending to the agricultural sector and thereby boost small-farm activities by providing guarantee to banks. The decree establishing the scheme initially provided for a Fund of N100 million subscribed to by the Federal Military Government of Nigeria (60 per cent) and the Central Bank of Nigeria (40 per cent). The share and called-up capital of the Fund has been raised to N3.0 billion

The objective of the Fund is to provide guarantee for of loans granted by any bank for agricultural purposes with the aim of increasing the level of bank credit to the agricultural sector.

The ACGSF is under the management of a body called the Agricultural Credit Guarantee Scheme Fund Board. The Central Bank of Nigeria is the Managing Agent charged with the day-to-day operations and administration of the Scheme. At the Head Office of the Central Bank of Nigeria, functions relating to the ACGSF are handled in the Development Department of the Central Bank. This department is headed by a Director. Any transactions respecting to the ACGSF in any state of the federation are carried out at the Central bank Office in the state.

The Agricultural purposes for which loans can be guaranteed by the ACGSF are those related to: establishment or management of plantation for the production of tree crops. Others are the cultivation or production of annual and biennial crops, animal husbandry, fish farming and fish capture, agricultural processing, farm machinery and hire services.

The maximum liability of the ACGSF with respect to any guarantee given under the Scheme is fixed from time to time by the Minister for Finance.

Currently, the liability of the Fund is 75% of the amount in default net of any amount realized by the bank from the security it got from the borrower. This is subject, in the case of a loan to an individual, to a maximum of N5, 000,000 and, in the case of a loan to a co-operative society or a corporate body, to a maximum of N10, 000,000.

The contributions of the Central Bank of Nigeria as a source of formal agricultural finance under the auspices of the ACGSF are substantial and impressive. In 2011 alone, loans guaranteed under the ACGSF were 5,829 in number and N1,390,579.55 in monetary value. From the year 1978 to the year 2010, the cumulative total of loans fully repaid were 502,227 valued at N25,425,415.95.

Innovations to the ACGSF by the Central Bank of Nigeria include Self-Help Group Linkage Banking, the Trust Fund Model and the Interest Drawback. Let us briefly look at these innovations as they relate to agricultural financing.

Self-Help Group Linkage Banking

The Self-Help Group Linkage Banking programme was inaugurated under the Agricultural Credit Guarantee Scheme Fund (ACGSF) in 1991 and started operation in 1992. Under the Self-Help Group Linkage Banking programme, farmers are advised to form groups of not less than five and not more than fifteen persons on the basis of common purpose. The groups are expected to undertake regular savings with a partner bank of their own choice. On diligently operating such savings account for not less than six months, the participating farmers as a group can then apply to the partner bank for loan. The total amount saved by the group provides part cash collateral for loan.

Loanable amounts to the groups are usually in multiples of the credit balance in the savings account as at the time of the loan application. The amount in the savings account which is a part of the security for the loan is not to be withdrawn from until the approved loan is fully repaid interest and principal. The Self-Help Group Linkage Banking is aimed at inculcating the culture and habit of savings and banking in the group members. It is also aimed at enabling the participating group members mobilize resources for financing their farm projects without recourse to bank borrowing on the long run.

The Trust Fund Model

The Trust Fund Model is a strategy for enhancing the supply of credit to the agricultural and rural sectors of the Nigerian economy. Under the Trust Fund Model, oil companies, State/Local Governments and Non Governmental Organizations (NGOs) place funds in trust with lending banks to increase and support the small group-savings of the farmers as collateral for agricultural loans.

Under the framework of the Trust Fund Model, the Trust Fund secures 25% or more of the intended loans of the prospective farmer-borrowers, the farmers' savings secure another 25% while the ACGSF guarantees 75% of the remaining 50%, thereby leaving the lending bank with a risk exposure of only 12.5%. In some cases, a state government, realizing the relatively low saving capacity of farmers, may decide to increase its stake beyond 25% in order to assist the smallholder farmers who may be unable to generate sufficient savings to qualify for a reasonable amount of loan.

Interest Draw Back Programme

The Interest Drawback Programme (IDP) was initiated as an interest rate management strategy under the ACGSF. The strategy is to reduce the effective borrowing rates without the problem of introducing dual interest rate regime or contravening the existing deregulation policy of the government. The IDP is instituted to provide support to borrowers for interest rate paid on loans under the ACGSF. Such support is at a rate referred to as the IDP rate which shall be determined by the Central Bank of Nigeria from time to time. As is constituted at present, the programme is for farmers borrowing under the ACGSF and is under the day-to-day operations and administration of the Central Bank of Nigeria.

Under the programme, farmers borrow from lending banks at market-determined rates. However, the programme provides interest rebate of a determined percentage to the borrowers where the loans are repaid as and when due. The IDP rate is announced by the monetary authority at the beginning of each financial year through the Monetary Credit Policy Guidelines and amended during the year as deemed necessary through the same medium, in line with movements of the Minimum Re-discount Rate (MRR).

The IDP rate is calculated on simple interest basis over the duration of the loan. The interest drawback shall only be available to the borrower through the borrower's bank account upon the eventual liquidation of the loan within the stipulated repayment period and for a maximum grace period of three months. The grace period does count for the calculation of the interest drawback entitlement.

For instance, a farmer borrowing N100,000 at 20% market rate for one year would pay N20,000 as interest to the lending bank. However, supposing the IDP rate at the time of borrowing was, say, 12%, that farmer, on liquidating the loan, shall be able to drawback N12,000. The farmer's effective borrowing rate has therefore been reduced by 12%.

The IDP has an authorized capital N2.0 billion. It is funded jointly by the Federal Government of Nigeria and the Central Bank of Nigeria in the ratio of 60:40. It is regarded as a dedicated fund for interest drawback on agricultural loans or IDP Fund and is separated from the ACGSF capital.

Other sources of agricultural financing under the auspices of the Central Bank of Nigeria but outside the ACGSCF include the following.

Agricultural Credit Support Scheme (ACSS)

This is an initiative of the Federal Government and the Central Bank of Nigeria with the active support, participation and collaboration of the Bankers' Committee. Bankers' Committee is a regular and compulsory meeting between the Central Bank of Nigeria (CBN) and all the managing directors of Deposit Money Banks operating in Nigeria. Usually, the Bankers' Committee is presided over by the CBN governor, in which participants discuss monetary policies as they affect the banking industry and the economy. The Scheme has a prescribed fund of N50.0billion and was introduced to enable farmers explore and exploit the untapped potentials of Nigeria's agricultural sector, reduce inflation, lower the cost of agricultural production, generate surplus for export, increase Nigeria's foreign earnings as well as diversify its revenue base. At national level, the scheme operates through a Central Implementation Committee (CIC) while at the Federal Capital Territory (FCT) and State levels, the Scheme operates through State Implementation Committees (SICs) instituted to ensure that the objectives of the scheme is realized without hitches.

To obtain loans under ACSS, applicants are advised to approach their banks for the loan through their respective state chapters of farmers associations and State Implementation Committees. However, large scale farmers are allowed under the scheme to apply directly to the banks in line with the guidelines.

ACSS funds are administered to farmers and agro-allied entrepreneurs at a single-digit interest rate of 8%. Under the scheme, banks grant loans to qualified applicants at 14.0% interest rate. Beneficiaries who pay back their facilities on schedule are to enjoy a rebate of 6.0 per cent, thus reducing the effective rate of interest paid by farmers to 8.0%.

Commercial Agriculture Credit Scheme (CACS)

As part of its developmental role in general and agricultural development in particular, the Central Bank of Nigeria (CBN) in collaboration with the Federal Ministry of Agriculture and Rural Development (FMA&RD) set up the Commercial Agriculture Credit Scheme (CACS) in 2009 (CBN, 2010). According to CBN (2016), from January 2015, 60.0 per cent of the Fund has been dedicated to the promotion of seven (7) focal commodities (rice, wheat, oil palm, fish, sugar, cotton and dairy) that contribute significantly to the nation's agricultural import bill. The scheme is aimed at providing funds for financing commercial agricultural enterprises. CBN (2014) defines commercial enterprise as any farm or agro-based venture with agricultural asset (excepting land) of not less than N100 million, for an integrated farm with potentials of growing the assets to N250 million within the next three years, and N50 million for non-integrated farms/agro-enterprise with capacity of growing the assets to N150 million. However, the definition makes provision for on-lending to farmers' cooperative societies.

According to CBN (2014), the scheme is financed from the proceeds of the N200billion three (3) year bond raised by the Debt Management Office (DMO). The scheme provides for each State Government in Nigeria to borrow up to N1.0billion for on-lending to farmers' cooperative societies and other areas of agricultural development whose activities are in line with the objectives of CACS.

The primary objectives of the Scheme are to:

1. Fast-track the development of the agricultural sector of the Nigerian economy by providing credit facilities to large-scale commercial farmers at a single digit interest rate;
2. Enhance national food security by increasing food supply and effecting lower agricultural produce and products prices, thereby promoting low food inflation;
3. Reduce the cost of credit in agricultural production to enable farmers to exploit the untapped potentials of the sector; and
4. Increase output, generate employment, diversify Nigeria's revenue base, raise the level of foreign exchange earnings and provide input for manufacturing and processing on a sustainable basis.

The Scheme which is a sub–component of the Federal Government of Nigeria's Commercial Agriculture Development Programme (CADP) is financed through the N200billion Bond raised by the Debt Management Office (DMO).

Under the scheme, the Central Bank of Nigeria requires all participating banks to sponsor projects from any of the target areas as contained in the guidelines and bear all the credit risk of the loans granted. Agricultural credit facility from the participating banks shall be in the form of loans. Loans to eligible entities under the Scheme are disbursed at a maximum interest of 9% inclusive of all charges. According to CBN (2017), this interest rate on CACS facility is shared between the participating bank and the CBN in the order of 7% and 2% respectively.

According to the CBN (2017), the Scheme shall be under the management of the Central Bank of Nigeria through the Board of Directors and the Committee of Governors. The Committee of Governors shall be responsible for the overall administration of the Scheme while Development Finance Department of the Central Bank of Nigeria shall be in charge of the day-to-day implementation of the Scheme.

On 15th August, 2017, the Central Bank of Nigeria stated that it had reviewed the guidelines of the CACS and therefore issued directive for the immediate commencement of insurance premium payments by borrowers under the CASC scheme. According to CBN (2017), the directive is to enhance the effectiveness of the CACS and further mitigate the risks faced by the participating financial institutions in financing the agricultural sector. The review and subsequent directive included a requirement that henceforth the Nigerian Agricultural Insurance Corporation (NAIC) should provide insurance cover for all agricultural facilities/projects under the CACS in line with the NAIC. Effective insurance cover is as incentive to agricultural production.

The Scheme is expected to terminate on September 30, 2025. This exit date does not apply to the tenor of individual loans and overdrafts which are based on their gestation period. According to the Central Bank of Nigeria (2017), as at 19th July, 2017, 513 projects have benefitted from the scheme with disbursement of over N473 billion.

Let us at this point look into other sources of formal finance for agricultural purposes.

Deposit Money Banks (DMBS)

Also known as commercial banks, or retail banks, they are dominant players in the financial intermediation process between the surplus and the deficit economic units of the economy. As at date, there are 22 deposit money banks in Nigeria. In varying degrees, they provide credit facilities to farmers. For instance, credit to the core private sector by the DMBs grew by 31.3 per cent at end-December2011, in contrast to the decline of 4.4 per cent recorded at end-December 2010. Of the amount outstanding, DMBs' credit to priority sectors constituted 36.1 per cent, of which3.5, 17.7, 0.5 and 14.4 per cent, were disbursed to agriculture.

Micro-Finance Bank

A microfinance bank is a bank that is tailored to extend small loans, also known as microloans, to individuals, businesses, and organizations that are categorized as low-income clients. It is a self-sustaining financial institution owned and managed by designated officials to render financial services to the targeted the low-income population where small amounts of money can go a long way. Because of the characteristic smallness of loans and savings, absence or reduced emphasis on collateral, and simplicity of operations, microfinance bank easily lends itself to the financing of the agricultural operations of the smallholder farmers. Microfinance bank in Nigeria provides the financial services to the low-income farmers. These financial services are either by commission or omission not provide to the farmers by the commercial banks in the urban areas. Microfinance bank in Nigeria generally operates in one location (Unit Microfinance Bank) as opposed to National Microfinance Bank which can operate in more than one state or State Microfinance Bank which can operate in one state of the Federal Capital Territory.

Bank of Agriculture (BoA)

In the beginning in 1973, the Nigerian Agricultural Bank Limited was established by the Federal Military Government of Nigeria to deal exclusively with agricultural loans.

The bank was meant to make funds available directly to cooperatives, credit-worthy individuals and governments in Nigeria. It was established to provide credit and loans for agricultural development projects. The name of the bank was changed to the Nigerian Agricultural Cooperative Bank (NACB) in 1978 with a renewed mandate to give loans to cooperative societies as on-lending and to individual farmers. The merger of such institutions as the Peoples' Bank of Nigeria and the Better Life for Rural Dwellers Scheme with the NACB gave birth to the Nigerian Agricultural Credit and Rural Development Bank (NACRDB) in the year 2000 to cater for financing and also integrate rural development activities.

The management of the Nigeria Agricultural Cooperative and Rural Development Bank (NACRDB) Ltd in 2011 announced a new name for the Bank. The institution is now known as the Bank of Agriculture Limited (BoA). The change of name follows the requisite administrative and regulatory approvals from supervisory agencies.

BoA is a Federal Government-owned development bank with a mandate to provide low cost credit to small-holder and commercial farmers, and small and medium rural enterprises. As a formal source of agricultural credit, Bank of Agriculture activity involves direct allocation of loans to farmers and the strengthening of unit micro finance banks, which deliver credit at the community level. BOA also embarks on various initiatives geared towards strengthening the cooperative credit structure at local and state government levels

BOA further provides micro financing to small and medium scale non-agricultural enterprises. It aims at ensuring effective delivery of agricultural and rural finance services on a sustainable basis to support the national economic development agenda, including food security, poverty reduction, employment generation, reduction in rural to urban migration, less dependency on imported food items, and increase in foreign exchange earnings.

Bank of Agriculture credit functions performed include direct lending, on-lending, collaboration, supervision and monitoring of credit. The functions involve: direct lending to qualified loan applicants engaged in agricultural and non agricultural small businesses; providing credit facilities to issuing institutions through on-lending service; and monitoring and supervising the flow of credit at the grassroots level.

The operational procedure of the BoA includes opening savings accounts by the applicants in any of the bank branches nationwide. The opened bank account must contain at least 20% of the amount of loan applied for. Applicants are requested to make a handwritten application for loan; a form is subsequently issued to the applicant. On approval, the loan amount received by applicants is dependent on the loan type. The following loan types are available: micro loans and macro loans. While micro loans range between N150000-N250000, macro loans range between N251000-N1m. Individual applicants get a maximum of N500000; corporate bodies have no stated upper limit.

Interest rate charged is negotiable and ranges between 12%-20% depending on the type of loan. For agricultural loans, the interest rate is12%; for non-agricultural loans, the interest rate is 20%. For macro loans the rate is 14%. The payback period of approved loans is dependent on the type of loan. For agricultural loans, it is 3 months; 1 month for non-agricultural loan and 6 months for macro loans. Safeguards against default include use of guarantors and constant monitoring of the loan.

In a study on the activities of Bank of Agriculture in Imo State, Nigeria, Oparah (2016), found that the amounts of loan disbursed by the Bank were rationed among numerous applicants. As a result, the recipients received low-sized inadequate loan amounts. For instance, while majority of the loan applicants represented by 45% applied for loan amounts amounting to the range N301000-N500000, majority represented by 65% obtained loan amounts ranging from N101000-N200000. No applicant applied for loan ranging from N51000-N100000 yet some 8% of the applicants were offered loan amounts in that range.

Nnamani and Mejeha (2017), in a study on evaluation of access to Bank of Agriculture credit by female farmers in Abia State, Nigeria, concluded that provision of credit by BoA to female farmers increased their farm production. Nnamani and Mejeha (2017), recommended the inclusion of extension agents in the lending institution's monitoring team.

The following section focuses attention on the informal sources of finance.

Informal Sources

Also called non-institutional sources, informal sources of loan include the following personal savings, friends and family, *isusu*, and money lenders. These sources are commonly characterized by direct and personal administration of loans from the lenders to the borrowers, relative ease and timeliness in the transaction, flexibility and variability of both repayment schedule and interest rates. Let us take a closer look at these non-institutional sources of finance.

Personal Savings

Personal savings refer to sums of money set aside by a person for a particular purpose including that of financing an agricultural venture. It can also be construed as money from a business owner's personal sources. Ordinarily, no repayment is required for personal savings

Friends and Family

This is usually a source of short term finance for an enterprise. Fund from this source can be secured or unsecured depending on the relation, agreement and the arrangement between the lender and the borrower.

Isusu

This is also called *esusu* or more glamorously rotating savings and credit association .It is a group of individuals who arrange to meet regularly for the purpose of saving and borrowing together. Basically, it refers to an informal financial intermediation arrangement in which a group of people come together to securely save and access their own money. Under this arrangement, *isusu* provides its members with microloans with or without interest rates depending on the prevailing arrangement. Lending by the *isusu* groups are usually deposit-based. Members of the group know themselves personally and so do not require collateral to secure the loans granted to members.

Money Lenders

Money lenders are traditional credit providers. The strength of the money lenders lies in the lack of bureaucracy, timelines and flexibility in credit terms. Money lenders do not ordinarily require a deposit relationship. Collateral may or may not be required; however, high interest rates are charged by the money lenders.

Semi-Formal Sources

Cooperatives

Cooperatives are a form of business organization where a group of people voluntarily organize to obtain, achieve or create a product or service. Cooperatives are broadly categorized as semiformal sources of agricultural finance to the extent that their activities are partially regulated by the government. It is a business outfit owned and democratically operated by the people who use its services and equitably share the costs and benefits of the business. A cooperative is a group of people acting together to meet the common needs and aspirations of its members, sharing ownership and making decisions democratically. It can be described as a business organization formed by the people and run by the people for the benefit of the people.

Cooperatives are to all intents and purposes based on the core cooperative values of self- help, democracy and equality, equity and solidarity. Self-help entails recognizing and acting on the need to provide for or helping to provide for specified help for oneself. Democracy and equality relate to membership as a human being with equal opportunity to vote and be voted for as opposed to undue rights and privileges. Democracy is at work to the extent that the cooperative business is owned and run by the members- the people who own and benefit from the cooperative services. Equity refers to the quality of being fair and impartial. Solidarity arises from the union or fellowship engendered by common responsibilities and interest of the members of the cooperative society who have come together to pool resources for specified purpose.

A cooperative is an autonomous association of persons who voluntarily cooperate and operate for the mutual social, economic and cultural benefit. For example, a cooperative can buy in bulk the inputs the members need to produce a certain product or service. The members then produce the product or service and then sells the good or service on behalf of the members. It should be stressed that a cooperative is not a charity organization but rather a distinct form of business organization that provides services or products to its members.

Co-operative societies are a source of financial intermediation. The main thrust of cooperative society's activities entails group mobilization and promotion of group action. Members' contributions are the main source of finance for cooperative societies. In addition to financial intermediation, other services which cooperative societies can render include input supply, and marketing of farm produce. Cooperative societies are a veritable organ for financial inclusion. Financial inclusion can be defined as the provision of a broad range of high quality financial services and products, such as savings, credit, insurance, payments and pensions, which are relevant, appropriate and affordable for the entire active population, and particularly the low-income group.

Co-operatives are a flexible business organization. They can be established in different ways, using different legal structures, depending on the vision and vision of the members. Although the cooperatives carry out different kinds of business, notwithstanding the flexibility of cooperative as a business organization, all cooperatives have core common characteristics. These include voluntary membership, basis, mutual or shared needs, cooperative principles, association of persons and joint ownership and democratic control of enterprise. Cooperative can be categorized into; consumers' cooperative; retailers' cooperative and producer cooperative.

Consumers' Cooperative

A consumers' cooperative is a business owned by the customers (consumers) of a particular good or service and managed democratically with the aim of fulfilling the needs of the members. This often takes the form of retail outlets owned and operated by the customers (consumers). Consumer cooperatives may sell consumer goods and services and the benefits are divided among the members based on their level of purchases. Consumers' cooperatives are run as a business as members vote on specific decisions and elect the members of board of directors from among the membership.

Retailers' Cooperative

Retailers' Cooperative is a form of business organization which employs economics of scale to receive discounts from manufacturers and markets the goods and services on behalf of the members. Ordinarily, economies of scale involve the increase in productivity arising from an expansion in the size of the market. In relation to retailers' cooperative, economies of scale work to the extent that when people pool their resources together and buy in bulk from manufacturers and thereby receive discount on account of the scale of purchase. The retailers' cooperative then sells at a reduced rate to the overall benefit of the members.

Producer Cooperative

Producer cooperative also known as workers cooperative is that form of cooperative owned and controlled by individuals who produce similar types of goods and services. Producer cooperative is common in the agriculture sector where farmers have to form cooperatives to survive in an economy that is increasingly industrial. By pooling resources and banding together, the cooperating producers (workers) increase and strengthen the extent to which they can enter into agreements that are favourable to them when discussing prices, terms of trade and other conditions with the buyers. In this form of cooperative, the cooperators also combine to more effectively market and brand their products.

Producer cooperative is wholly and entirely owned by the workers themselves as no "outsider owner" ownership is allowed. Membership to producer cooperative is characteristically voluntary. Majority of its work force owns shares, and the majority of shares are owned by the workforce.

According to CBN/ RUFIN/ UNDP (2010), citing the 2002 National Policy on Cooperative Development, the weaknesses of Farmers Associations including cooperatives are as summarized:

1. The general low level of literacy and awareness among members of Farmers Associations;
2. Limited skill and capacity to operate outside their immediate environments;
3. Lack of transparency and accountability among leaders of Farmers Associations leading to high incidence of fraud and mismanagement of funds;
4. High level of mortality among Farmers Associations that are formed to target specific government or donor benefits;
5. Dependency syndrome associated with excessive government pampering, interest and involvement;
6. Small average size of (farmers) cooperatives;
7. Paucity of capital;
8. Apexes which are not viable, and which do not render services to their affiliates;
9. Inability to draw primary producers into cooperatives;
10. Weak internal linkages;
11. Lack of government agents in the distribution of credit and agricultural inputs;
12. Low rate of loan repayment due to the 'national cake' syndrome; and
13. Inability to overcome socio-cultural barriers to achieve national spread.

Let us investigate a government-backed intervention programme that made appreciable attempt at forging the activities of formal, informal and semi-formal into an arrangement for achieving an enabling microfinance environment for rural poverty reduction and enhance rural poor's access to financial services in order to expand and improve agricultural productivity and Micro-Small Rural Enterprises in Nigeria.

RURAL FINANCE INSTITUTION-BUILDING PROGRAMME

Access to financial services especially to the small-scale farmers is critically important to the foundations of economic development. It is often debated as to what constitutes access: is it availability of financial services for people in a given location for a particular period or is it actual and profitable utilization of the services by the clientele for which the financial services are meant to serve? It should be stressed that access to financial services can be construed from the points of view of both availability and actual utilization in a beneficial way. Access to financial services especially for the purpose of securing credit facilities for funding of enterprises also entails opening and maintaining savings accounts in a financial institution. For the rural agricultural enterprises and the associated small holder farmers, microfinance institutions are tailor made for such financial services. CBN/ RUFIN/ UNDP (2010), is of the view that while much attention has been focused on access to credit, very little emphasis has been placed on what influences individuals in Nigeria to open and maintain savings accounts. Savings accounts are likely the first and most obvious way in which household participation in the formal financial sector begins; additionally, savings especially financial savings are the commonest source of funds for starting a business.

According to CBN/ RUFIN/ UNDP (2010), there have been four major attempts to link cooperatives (formal and informal) with the banking system. The first was a small pilot effort in present Ondo state between 1968 and 1971 building on the foundations of pre-existing *esusu*. The second occurred in the 1980s and involved self-help groups (SHGs) organized to borrow from commercial banks at controlled interest rates and guaranteed under the Agricultural Credit Guarantee Scheme Fund (ACGSF) of the Central Bank of Nigeria. The third trial was also launched by the CBN in the early 1990s and served 137 cooperatives and 176 informal groups. However, this effort though supported by compulsory lending requirements on commercial banks, failed when new low interest caps were imposed in 1994. Since the year 1994, there have been some other attempts, either by state governments or the Central Bank of Nigeria with varying degrees of success.

The Rural Finance Institution Building Programme (RUFIN) is a Loan Agreement between the International Fund for Agricultural Development (IFAD) and the Federal Government of Nigeria. According to CBN/ RUFIN/ UNDP (2010), Rural Finance Institution-Building Programme (RUFIN) is being supported by the International Fund for Agricultural Development (IFAD) to effectively and efficiently direct policies to sustainably service the agricultural and rural sector, strengthen institutions that will bring financial services to the poor and facilitate the participation and inclusion of rural people and their institutions into the national financial system.

RUFIN was approved by IFAD's Executive Board in September 2006 and became effective on 20 January 2010. The completion date is January 2017.

Total programme cost is USD 40 million, of which IFAD is contributing a loan of USD 27.18 million and a grant of USD 0.4 million.

Aims and Objectives of the Rural Finance Institution Building Programme

According to IFAD (2013), the overall development goal of RUFIN is to create an enabling microfinance environment for rural poverty reduction. Specifically, the Finance Institution-Building Programme (RUFIN) aims at enhancing access of rural poor households, particularly woman headed households, physically challenged, and rural youth to financial services. The programme aims to develop and strengthen Micro Finance Banks (MFBs), other member-based Micro Finance Institutions (MFls), by enhancing the access of the poor rural dwellers to the financial services of these institutions and thereby expand and improve agricultural productivity and Micro-Small Rural Enterprises. The objective is to develop and strengthen microfinance institutions (MFIs) and establish linkage between these institutions and formal finance institutions, in order to create a viable and sustainable rural finance system.

According to CBN/ RUFIN/ UNDP (2010),

The program is funded by IFAD and the Federal and State Governments of Nigeria (FGN). It focuses on building the capacity and strengthening Rural Finance Institutions (RFIs), namely rural MFBs, rural finance cooperatives/ NGOs and member-based rural group organizations to enhance their effectiveness and be able to facilitate access to financial services to the rural poor, assisting with necessary policy changes, promoting sector initiatives to create conditions for rapid private sector-led poverty reduction and economic growth; and improving accountability and transparency.

The four key partners of the programme at the National level are (a) Nigeria Agricultural Cooperative and Rural Development Bank, now renamed Bank of Agriculture, BoA, (b) Central Bank of Nigeria, (c) National Programme for Agriculture and Food Security, and (d) Federal Department of Cooperatives (FDC) of Federal Ministry of Agriculture & Rural Development. The National Programme for Agriculture and Food Security (NPAFS) is the implementing Agency.

Target Group and Area of Coverage

The Rural Finance Institution-Building Programme target group is made up of women, youth, physically challenged and farmers. Rural poor households, particularly female-headed households are particularly targeted by the programme. The target groups were expected to benefit directly from improved financial services in terms of quality and quantity, through access to deposit services, loans and payment or transfer services. About 345,000 families (of which at least 138,000 [40%] will be female-headed households) were estimated to benefit directly from the programme.

The seven-year RUFIN programme was designed to cover two states in each of the six geo-political zones. The programme effectively covered 12 states of Nigeria. Three local governments were selected in each of the twelve states. The programme covered 36 Local Government Areas (LGAs). According to CBN/ RUFIN/ UNDP (2010), In order to ease implementation, the project was divided into three zones with a coordination arrangement for each zone, and an overall coordination arrangement for the project. The Project Zones are:

1. The Northern Project Zone comprising the Northeast and Northwest geo-political zones
2. The Middle-belt Project Zone comprising North Central and South West geo-political zones
3. The Southern Project Zone comprises South East and South-South geo-political zones.

The 12 states of the six geopolitical zones of Nigeria covered by the RUFIN programme are: the Northwest (Katsina and Zamfara States), Northeast (Adamawa and Bauchi States), North Central (Benue and Nasarawa States), South-West (Lagos and Oyo States), South-south (Akwa-Ibom and Edo States) and Southeast (Anambra and Imo States).

The Rural Finance Institution Building Programme (Rufin) Components

The Rural Finance Institution Building Programme (RUFIN) had the following components and subcomponents:

- **Component 1:** Capacity Building and Technical Support to Bank and Non-bank Micro-finance Institutions:
- **Subcomponent 1.1:** Capacity Building of Bank and Non-Bank Financial Institutions
- **Subcomponent 1.2:** Promotion of Access to Refinance and Linkage facilities.
- **Component 2:** Targeted Development and Strengthening Institutional Environment for Microfinance Development
- **Subcomponent 2.1:** Improvement of the policy, legal and institutional framework for cooperatives and Rural Micro Finance Institutions;
- **Subcomponent 2.2:** Implementation support, regulation and supervision of Micro Finance Banks; and
- **Subcomponent 2.3:** Research and documentation of policy dialogue.
- **Component 3:** Programme Coordination and Management.
- **Subcomponent 3.1:** Project management unit
- **Subcomponent 3.2:** Monitoring and evaluation (M&E)
- **Subcomponent 3.3:** Policy support for implementation.

Operational Procedure of the Rural Finance Institution Building

Under the auspices the programme, such Non-Bank Micro Finance Institutions as Cooperative Societies, Unions and Cooperative Finance Agencies (CFAs) and Grassroots informal Finance Institutions were supported in the needed areas of capacity building and access to loanable funds. The support is ultimately aimed establishing appropriate, effective and sustainable linkages with micro finance banks/ institutions and commercial banks for credit delivery to the rural populace. It is managed through a Central Programme Management Unit (CPMU) in Abuja, Nigeria. The CPMU is headed by the National Programme Coordinator (NPC) who is assisted by a set of managerial and support staff members. According to (IFAD, 2012), the Central Programme Management Unit implemented the programme through three Zonal Offices, 12 State Offices, and 36 LGA level establishments. RUFIN's overall implementation procedure can be grouped into three tiers (a) the bottom tier dealing with rural finance market development at the community level. This tier is charged with developing financial awareness, stronger group management, savings and credit culture; (b) the middle tier, is saddled with the strengthening of the supply of rural financial services through Microfinance Banks (MFBs), non-bank financial NGOs and apex cooperatives; and (c) the top tier which is concerned with the creation of the enabling environment, funds flow and sustainability support to the sector. IFAD (2012), further stated RUFIN was designed to pursue the flow of funds into rural areas by influencing different stakeholders but, without directly being involved in the credit delivery chain.

Assessment of the Rural Finance Institution Building Programme

IFAD (2015), an overall assessment of the Rural Finance Institution Building Programme implementation under the bottom tier, the middle tier and the top tier as follows:

Bottom Tier

As at 30 September 2015, RUFIN had mentored 10,550 groups of which 3,000 were pre-existing groups strengthened by the programme and the remaining 7,550 were mainly women's groups formed since late 2013 with strong focus on internal savings and credit practices ((IFAD, 2015). IFAD (2015), further identified two particular changes attributable to the RUFIN initiative: (a) increasing adoption of internal savings (and credit) as frontline group activity as against groups being driven solely by external credit linkages; and (b) external credit linkages increasingly based on business strategy of MFBs/MFIs instead of being obligated by RUFIN support received by the MFBs/MFIs. Success stories of women's groups have increased as they have been faster in adopting the savings led approach, accumulating funds and accessing internal credit at 10-15% compared to at least 30% charged annually by external lenders.

According to IFAD (2015), the main concern at the bottom tier is the very low density of groups especially in the far away rural communities with sometimes only 1-2 very successful groups present in a community of greater than 1,000 households. This tends to diminish programme impact and affects cost efficiency of potential financial institutions willing to increase rural outreach through the group approach. IFAD (2015), recommended increasing group formation either directly by RUFIN state and local government staff or indirectly through volunteers and RUFIN mentored MFIs (R-MFIs), in those villages where R-MFIs have already started group linkages. While IFAD (2015) was concerned about the very low density of groups especially in the out of the way communities, World Bank (2008), had raised a concern on the quality of group formation as a basis for credit delivery. As an alternative approach for specifically targeting the less bankable rural poor World Bank (2008), recommended actively promote savings groups, with emphasis on the quality of group formation. It makes more sense to worry more about the quality than the density of the groups formed. Group quality can be mirrored from member participation and group solidarity, good organizational support and social mobilization. Groups formed in direst reaction to immediate incentives (such as accessing government loans) tend to disband once the members' short-term goal is accomplished (World Bank, 2008 CGAP, 2007; Fouillet, 2007). To all intents and purposes, emphasis should be placed on voluntary and high-quality group formation.

Middle Tier

RUFIN had successfully focused the middle tier partners on the rural outreach objective and has guided them to adopt group methodology and savings led approach for achieving rural outreach (IFAD, 2015). According to the IFAD (2015), compared to an earlier situation when RUFIN-mentored Microfinance Institutions (R-MFIs) viewed groups as only a means for delivering credit, the R-MFIs now consider groups as veritable sources of savings first. Furthermore, 245 MFB/MFIs from the 12 RUFIN States had developed Rural Business Plans (RBPs). Several R-MFIs have already commenced group formation and savings mobilisation according to RBP targets.

Start-up implementation challenges have emerged. As R-MFIs were not trained on group development their approach in this area is not standardised. Staff shortage and rural transportation challenges are the other obstacles along with gaps in start-up portfolio funds and capacity. Rural transportation challenges as they relate to rural development have been highlighted elsewhere. FMARD (2016) had highlighted the need to deepen the supporting infrastructure to reduce poverty in rural areas, alleviating the suffering of rural dwellers and creating enablers for economic take-off in the rural areas. Government, in partnership with relevant stakeholders can achieve this through the systematic provision of individual infrastructural facilities and also through the integrated approach to rural development.

Top Tier

According to IFAD (2015), the successful roll out of the Rural Business Plans training through National Association of Microfinance Banks and Association of Non-Bank Microfinance Institutions has improved their capacity to partner with projects. The State Rural Outreach Coordination Committees (ROCCs), have been mobilized with regular meetings and active participation by members. Some RUFIN mentored Micro Finance Institutions have accessed and delivered funds to RUFIN groups.

Although the CBN has adopted the rural outreach mandate, it is yet to develop an internal package of interventions for delivering on this agenda on a wider scale. This is a concern for sustainability. Although the ROCCs have made a promising start at the State level it is currently unclear how their actions from RBP implementation will seamlessly converge and influence policy and enabling environment decisions at the National level.

According to CBN/ RUFIN/ UNDP (2010), there has been a mosaic of institutions, programmes, governmental, non-governmental and a combination of both governmental and non-governmental implemented in Nigeria to address poverty. These different institutions and programmes include many of those which involve the provision of credit often on subsidized basis with elements of targeting. Regrettably, these interventionist institutions and programmes have not been as successful as hoped either in absolute terms or in comparison to best practices in other countries. The amount of tax payers' money committed to these institutions and programmes is simply huge. Their impact either in the creation of sustainable financial institutions serving the poor and financially excluded or in impact on poverty has again not been as successful as expected.

In the views of World Bank (2008), there is a legacy of lack of financial sustainability and operational inefficiency of most government and many non-governmental rural finance institutions and programmes. Successful future interventions in rural finance must take place in the context of continuous strengthening of private and public institutions.

SOLUTIONS AND RECOMMENDATIONS

It is widely reported that banks are reluctant to lend to agriculture. This reluctance is largely dictated by the profit motives of banks against the inherently risky nature of agricultural enterprises. A way round the problem of bank's unwillingness to lend to agricultural enterprises is through efficient government backed and well-targeted credit guarantee schemes. The Agricultural Credit Guarantee Scheme Fund of the Central Bank of Nigeria is therefore commended and should be made more efficient.

Farmers on their part are advised to form cooperative societies and obtain loans from such cooperative societies at single digit interest rates. The government should constantly and consciously discharge its responsibilities as they relate to supervision, monitoring and evaluation of the activities of registered cooperative societies.

Non-governmental organizations who may consciously decide to invest in the provision of credit facilities to farmers should be provided with the enabling environment. Such enabling environment can include tax relief to the extent of the investment and also provision of necessary regulatory framework for contract enforcement.

Nigerian government should strive to commit itself to the Comprehensive Africa Agriculture Development Program (CAADP) call for the allocation of at least 10% of the national budget for agriculture to achieve a target of 6% annual agricultural growth.

Institutional, social and technical sustainability of interventions and programmes depend on the continued implementation, by the direct and immediate and not so direct and not so immediate beneficiaries of the relevant interventions and programmes, of the lessons learnt beyond the closure of the programme. With particular reference to the Rural Finance Institution Building Programme (RUFIN) Institutional sustainability depends on continued implementation of the rural outreach mandate led the CBN, in RUFIN states and beyond after RUFIN's closure. This can be achieved by government taking the lead in creating the enabling environment and good governance. In this sense, the concept of integrated rural development should be given greater impetus, first by the government and then by the government in partnership with relevant private stakeholders.

FUTURE RESEARCH DIRECTIONS

Formal financial institutions tend to be reluctant to lend to the agricultural sector. Informal financial institutions tend to lend to agriculture at somewhat usurious interest rate. Semiformal financial institutions such as cooperative societies have been traditionally associated with the function of pooling financial resources of members for mutual benefit. Research is recommended on the role of cooperatives in effectively and efficiently financing agricultural enterprises in Nigeria.

There has been a mosaic of programmes, governmental, non-governmental and a combination of both governmental and non-governmental implemented in Nigeria to address poverty. These different programmes include many of those which involve the provision of credit often on subsidized basis with elements of targeting. Regrettably, these interventionist programmes have not been as successful as hoped either in absolute terms or in comparison to best practices in other countries. The amount of tax payers' money committed to these and programmes is simply mind boggling. Their impact either in the creation of sustainable financial institutions serving the poor and financially excluded or in impact on poverty has again not been as successful as expected. It is hereby suggested that an empirical study be conducted on the sustainability strategies of credit and savings programmes interventions in Nigeria in the past ten years.

CONCLUSION

Funds for financing agricultural enterprises have been traditionally obtained from both formal and informal sources. The formal sources are capable of mobilizing a large of money, however, in the process of financial intermediation, a relatively small fraction of the whole gets to agriculture. The informal sources tend to mobilize funds for agriculture at high and therefore unsustainable interest rates to the farmers. Incentives and subsidies are therefore required for financing agriculture. A typical example of incentive is effective insurance cover. Relevant stakeholders including governments and non-governmental organizations should be actively involved in the incentives and subsidies. In the case of institutional interventionist programmes, sustainability should be clearly worked out by the relevant authorities before tax payers' money is committed.

REFERENCES

Central Bank of Nigeria. Rural Finance Institution Building Programme, & United Nations Development Programme. (2010). *Report of UNDP/CBN / RUFIN baseline survey of the microfinance sector*. Retrieved August 13, 2017, from http://0837CBN_Report_0f_Microfinance_Survey_2010

Central Bank of Nigeria. (2014). *Guidelines for Commercial Agriculture Credit Scheme (CACS) Central Bank of Nigeria (CBN) and Federal Government of Nigeria*. Central Bank of Nigeria.

Central Bank of Nigeria. (2016). *Monetary, Credit, Foreign Trade and Exchange Policy Guidelines for Fiscal Years 2016/2017 (Monetary Policy Circular No. 41)*. Central Bank of Nigeria.

Central Bank of Nigeria. (2017b). *Commercial Agriculture Credit Scheme Guidelines*. Central Bank of Nigeria.

Central Bank of Nigeria. (2017a). *CBN disburses N473bn under Commercial Agricultural Credit Scheme*. Retrieved October 18, 2017, from http://insidebusinessonline.com/index.php/2017/07/19/cbn-disburses-n473bn-commercial-agricultural-credit-scheme/

Central Bank of Nigeria (CBN). (2010). *Guidelines for Commercial Agriculture Credit Scheme (CACS) Central Bank of Nigeria (CBN) and Federal Ministry of Agriculture and Rural Development (FMA&RD)*. Central Bank of Nigeria.

CGAP. (2007). *Sustainability of Self-Help Groups in India: Two Analyses*. Occasional Paper No.12. Washington, DC: IBRD.

Federal Ministry of Agriculture & Rural Development (FMARD). (2016). *The Agriculture Promotion Policy (2016 – 2020) Building on the Successes of the ATA, Closing Key Gaps Policy and Strategy Document*. Abuja, Nigeria: FMARD.

Fouillet, C. (2007). Spread of the SHG Banking Linkage Program in India. In *International Conference on Rural Finance Research: Moving Results*. FAO.

International Fund for Agricultural Development (IFAD). (2012). *Rural Finance Institutions Building Programme (RUFIN) 4th Supervision Mission: Supervision Report – Aide Memoire, Appendices and Annexes (Mission 11-26 June 2012)*. IFAD.

International Fund for Agricultural Development (IFAD). (2013). *Rural Finance Institution Building Programme (RUFIN) Mid-Term Review Report*. IFAD, West and Central Africa Division.

International Fund for Agricultural Development (IFAD). (2015). *Federal Government of Nigeria Rural Finance Institution Building Programme Supervision report*. IFAD West and Central Africa Division Programme Management Department.

Nnamani, N. G., & Mejeha, R. O. (2017, October). *Evaluation of access to credit from bank of agriculture (boa) for improving agricultural production among female farmers in Abia State, Nigeria.)*. Paper presented at the Faculty of Agriculture, Federal University Ndufu Aike Ikwo, Ebonyi State, Nigeria First International Conference on Food Security and Hidden Hunger University Auditorium Ebony State Nigeria.

Oparah, C. M. (2016). *Assessment of the outreach of bank of agriculture in Imo State, Nigeria* (Unpublished bachelor degree dissertation). Imo State University, Owerri, Nigeria.

World Bank. (2008). *Rural Finance in Nigeria Integrating New Approaches*. Agriculture and Rural Development Unit Sustainable Development Network Western Africa Country Department Africa Regional Office.

Chapter 3
Economics of Farm Management

ABSTRACT

The concept of farm management can be defined as the organization of land, labour, capital, and management (decision-making and risk-taking) in operating a farm enterprise to meet the stated objectives including profit maximisation. Of the four fundamental factors of production, management stands out in its onerous role in organizing all the others in meeting the defined objectives of the enterprise. Onerous as the role of management as a factor of production is, it cannot work in isolation of the others. There is therefore a need to discuss farm management as a concept with due focus on the notion of entrepreneurial spirit and at the same time bring out the peculiar features of the factors of production in working together to maintain a going concern. It is against this background that this chapter discusses the concept of farm management. The specific objectives of the chapter include defining the meaning and scope of farm management, highlighting the resources available to the farm enterprise, highlighting the relevance of entrepreneurship behaviour and entrepreneurship spirit in growing an enterprise, and highlighting the importance of farm management. The methodology adopted is descriptive. It is argued in this chapter that present and prospective farm managers should essentially have entrepreneurial spirit with which to start, grow, and expand their enterprise in line with the objective of the enterprise in question. Searching out investment opportunities and effectively ploughing back profits on existing or new investments for expansion enterprise is one of the decisions of management.

DOI: 10.4018/978-1-5225-3059-6.ch003

INTRODUCTION

This chapter is focused on the concept of farm management defined as the organization of land, labour, capital and management (decision-making and risk-taking) in operating a farm enterprise to meet the stated objectives including profit maximisation. The specific objectives of the chapter include: define the meaning and scope of farm management, highlight the resources available to the farm enterprise, highlight the relevance of entrepreneurship behaviour and entrepreneurship spirit in growing an enterprise, and highlight the importance of farm management. The chapter is concluded with recommendations.

MEANING AND SCOPE OF FARM MANAGEMENT

Management involves the control and making of decisions in a business or similar organization. The complete range of farm business management refers to the process of making and implementing decisions involved in planning, organizing and operating a farm business for optimum productivity and attendant profit. Productivity refers to the ratio of the amount of output per unit of input employed in the production process.

The science and art of farm management borrows from the knowledge of such disciplines as production economics, marketing, and finance including government policy pronouncements. Farm management is concerned with the organization of land, labour, capital and human resources in operating a farm enterprise to meet the stated objectives. It deals with the profitability and profitable organization of individual farm units. Farm management entails control and decision-making processes in a given farm unit. It also involves actual execution of decision taken. The individual at the centre of the decision-making process is the farm manager.

The resources of land, labour, capital and human beings available to the farm firm for management and ultimate production of the required goods and services are collectively called productive resources or factors of production. Availability and efficient control and decisions over these resources go a long way in defining the attainment of the stated objectives of the farm enterprise. Let us take a closer look at these resources available to the farm firm.

RESOURCES AVAILABLE TO THE FARM ENTERPRISE

Land

In terms of farming activities, land is the single most important factor of production. Agricultural production cannot ordinarily take place without land. It can be regarded as the sum total of natural and man-made resources over which the possession of earth surface gives control (Fabiyi, 1979). As a factor of production, land can be defined as the totality of the original and indestructible properties of the earth's crust; it refers to all natural and naturally occurring resources other than human beings and they include soil, water, rainfall, sunshine, temperature and air.

With respect to land mass, Nigeria has a land mass that spans over 91.08 million hectares out of which 36.50 million are for arable crops, 3.00 million for permanent crops and 39.00 million for pasture (FAOSTAT, 2010). It should be stated that Nigeria land mass is under the constant stress of unmitigated environmental challenges and disasters. These disasters and challenges directly threaten the peoples' lives and livelihoods. Desertification threatens not less than 11 States of Nigeria. Gully Erosion is a perennial problem in several states in the South East, South South and the North Central. Coastal Erosion affects the whole national coastline.

In many locations in Nigeria, erosion, an advanced form of soil degradation had resulted to gully erosion. In varying degrees, gully erosion in. the different locations had deteriorated through the combined effect of landslides. With more than 2000 active gully erosion sites across Nigeria, coastal erosion annually wears away 2-30 metres of land. Gully erosion is prevalent in the Southeast and South-south, and North-Central zones; coastal erosion threatens the tidal basins, beaches and creeks. In the north, wind erosion predominates as an environmental problem.

Land is ordinarily held individually in families and communities in Nigeria. Acquisition of land for farming can be effected either through the individual owners and/or the community leaders. Acquisition can be arranged under owner operator, rent and outright purchase.

For the small-scale farmers in general and the young people in particular, trying to make a livelihood in agriculture, access to land is critically important. According to FAO (2014), access to land is not only the number one requirement for starting farming, but it can also contribute to household food security and is a means for employment creation and income generation. Access to land serves as security and collateral for accessing credit. Access to land should be qualified to imply sustainable and secure access for efficiency and effectiveness in farm management.

The critical importance of land as a factor of production notwithstanding, many small-scale farmers do not have the required sustained and secure access. This is can be attributed to the traditional land tenure system operative in the different communities. In Nigeria and in most developing countries, the common and popular way for accessing land is through inheritance. With the population growth trending upwards and the supply of land fixed, access to land tends to be fraught with challenges. The challenges faced by men in accessing land are multiplied for women especially young and unmarried women.

In some developing countries, women tend not to inherit land and only obtain user rights via a male relative. In some countries, this type of land tenure system has been reformed so that women are granted equal property and inheritance rights with men. For instance, the Land Use Act Chapter 202 Laws of the Federation of Nigeria 1990, Part II section 5(1) a states that it shall be lawful for the Governor in respect of land, whether or not in an urban area to grant statutory rights of occupancy to any person for all purposes. However, the enforcement of such formal laws can be an uphill task; also parallel customary law systems operate and tend to deny equal land access for women. The extant Land Use Act is particularly due for reform.

According to Oshio (1990), prior to the introduction of uniform rights of occupancy system, Nigeria operated a plural system of land tenure namely: tenure under the received English law; tenure under the State Land Laws; tenure under the Land Tenure Law; and the indigenous tenure under customary law. English law, which was received into Nigeria as part of the political process, created certain interests in land and governed tenure in these interests nationwide.

Tenure with respect to the parcels of land all over the country which were specifically taken over by the state was governed by the provisions of the various State Land Laws. The customary tenure in northern Nigeria experienced early disruptions by the Fulani jihadists, who introduced a somewhat feudal tenure under which they claimed over-lordship of the land after the Islamic conquest. When Nigeria became a colony of Britain, the leadership of Lord Lugard, to whom land rights were ceded in 1903, introduced statutory regulation of land rights under the Lands and Native Rights Ordinances of 1910, as amended in 1916. According to Oshio (1990), the 1916 Ordinance was amended and reenacted in the Land Tenure Law of 1962. This law designated certain lands in northern Nigeria as "native lands" and vested their management and control in the Minister (later Commissioner) for Lands and Survey to administer for the use and common benefit of the natives. The consent and approval of the minister was also required for the occupation and enjoyment of land rights by non-natives (Oshio, 1990).

Tenure in southern Nigeria as regulated by customary law had its roots in the traditional conception of land as having economic, social, political, and religious significance; it was conceived of as a sacred institution given by God for the sustenance of all members of the community, and as such it belonged to the dead, the living, and the unborn. The concept of community/communal ownership of land should be understood in its fundamental sense. For instance, when an individual wants to buy or sell a piece of land a community of relevant people is essentially involved in the deal as witnesses. Under the customary law land tenure system, land belongs to the community, the village or the family, never to the individual.

According to Oshio, (1990), although there is no direct reference to the indigenous land tenure in the Land Use Act, the recognition and preservation of customary land law within the language of the Act may imply the survival of the indigenous land tenure. For example, Section 24 preserves the customary law rules governing devolution of property, while section 25, which prohibits partitioning of land, and expressly exempts cases which are regulated by customary law.

Oshio (1990), concluded that based on experience, the letters of a statute are not inherently sacred and that their ability to accomplish the desired goals depends largely on the operators. The Land Use Act is no less subject to this idea. Oshio (1990), observed that although the Act seems to recognize the indigenous land tenure, it nonetheless fails to make clear and adequate provisions for it. For instance, section 48 saves existing laws subject to modifications, but until now the extent of these modifications has not been made clear. Accordingly, there is no doubt that there is need for reform. Two alternative reforms are suggested. The first is an amendment of the Land Use Act to make clearer provisions for the indigenous land tenure system so as not to leave everything to mere implication, as is presently the case under the transitional provisions of sections 34 and 36. The alternative suggestion is an amendment of the Act to exclude the indigenous land tenure system from the operation of the Act. This would enable the latter to co-exist with the rights of occupancy system without the present problems of interpretation.

These challenges in addition to the fixed nature of supply of land especially in the short term, confront farmers and farm managers with high cost of renting and/or purchase of land. The challenges also have the tendency to bring about increased fragmentation of land owned through inheritance.

It has been proposed that loans should be made available to farmers to assist them have access and/or ownership of land. While loan may be granted for defraying the fixed cost of production of which rent is an integral part, the small-scale farmer may find it difficult to secure loan to enable him own land.

Labour

Labour is another factor of production. It can be defined as the amount of work done. Alimi, (2012) defined farm labour as the quality and quantity of human efforts available for farm work. Labour refers to all human activities that provide the goods or services in an economy. It includes both physical and mental labour. For instance, the activities of workers who weed a farm are a form of labour just as the activities of service providers who operate machines or tend to work animals. Labour is a human factor of production whose services are rewarded with wages. This is contrasted with the services rendered by entrepreneurs (the risk-bearers of the enterprise) which are rewarded by profits. It can also be defined as all inherited, acquired, mental and physical human resources involved in a production process. Labour can be obtained from human beings, machines and animals.

Farm labour can be broadly classified into family labour and hired labour. Family labour includes work done in a farm by the household members and may or may not be paid for. Family labour is the labour supplied by the farming family, which consists of the man hours made available by the farmer, his wife or wives, children and relations. Labour from family sources is ordinarily restricted by the number of able bodied and active members of the family. It is recommended that children should not be employed in agricultural production to the extent that their formal educational pursuits are jeopardized. Where children have to be employed, it should be outside the school hours. It is argued that changes in family size, free and compulsory education for children, and rural to urban migration of youths tend to affect the supply of labour from family labour source.

Hired labour, to all intents and purposes are paid for. External injection of capital is necessary to enable small-scale farmers pay for hired labour. Payment for hired labour is determined by the forces of demand and supply. The supply of hired labour is constrained by its high cost. Small-scale agriculture is characterized by labour- intensive practices.

The amount of labour used to accomplish a given task depends on the nature and size of the task in question and also on the wage rate. It should be emphasized that labour refers to the amount of work done by human beings per unit of time. It is therefore measured as a flow over a given period of time as opposed to a stock which refers to point in time. Labour inputs are measured in man-hours. Man-hour is the product of the number of men employed and the average number of hours worked.

Capital

Capital, another productive resource, refers to human-made aids to further production including tools, machines, buildings which are used up in the course of producing other goods as opposed to being consumed for their own sake. It refers to total investment available to the farm firm for use in generating the desired output. Capital includes simple tools, machinery, seeds, fertilizers, chemical, livestock and improvements on land.

It should be stated that money is used to finance the acquisition of the productive resources and as a result should strictly not be categorized as a factor of production. Access to these productive resources is enhanced by the amount of financial resources available to the farm manager. This is where cash/money comes in discussing capital as a factor of production.

Access to financial services such as savings, loans, credit, and insurance is of fundamental importance to start and financially sustain any agricultural activity. Access to and efficient utilization of financial services enable farmers and farm managers to finance the costs of planting and harvesting, as well as investments in improved productive capacities. Financial service providers play a crucial role in financing capital investments in agriculture.

The financial service providers include the commercial and development banks, semiformal banking system such as savings and credit cooperative organizations and informal banking systems not officially registered at national level e.g. self-help groups village savings and loan associations. Most financial service providers tend to focus more on credit than on the provision of savings and insurance services. Even where institutions for settling insurance claims of farmers exist, the performance and activities of these institutions are anything but satisfactory.

The financial service providers, in the exercise of their legitimate discretion and due diligence request clients to provide collaterals before funds are released. Items used as collaterals provide security to the lender, and help ensure that the lender gets his or her money back even if the borrower is unable or unwilling to satisfactorily repay the loan in full and in due time. The collaterals can be in the form of a borrower's pledge of a thing of value, for example property, certificates etc, to the lender in a bid to secure the repayment of the loan. Other forms of collaterals include surety, guarantor and cash. Formal banks require collaterals which the smallholder farmers find difficult to provide. Such difficulty may be avoided through group lending where members of a group of 5-10 come together to access a loan facility and thereby guarantee one another.

Providing financial services to the agricultural sector is typically considered to be risky due to the unique characteristics of agriculture: dependence on natural resources and seasonality; long production cycles; and vulnerability to variable weather (FAO, 2014). Let us examine the concepts of business and financial risks in a production enterprise.

Concepts of Business and Financial Risks

Business risk is the risk inherent in the firm, independent of the way it is financed. Business risk is generally reflected in the variability of net cash. A high (low) coefficient of net cash flows, for example, would indicate high (low) business risk.

There are two major external sources of business risk in the agricultural enterprise: the market which produces variability in price for both outputs and inputs and the uncertainty in the availability and the quality of the latter; and the biophysical environment which produces yield variability. The level of business risk is also influenced by internal factors such as investment decisions and management skills.

Some approaches to business risk modification include insurance, government programmes, whether modifications and innovations of individual farmers. Introduction or modification of risk in the production process affects the pattern of resource allocation and in turn the level of production.

Financial risk is the added variability of the net cash flows of the owners of equity that results from the fixed financial obligation associated with debt financing and cash leasing. Financial risk encompasses the risk of cash insolvency.

There is a financial response to business risk modification as business risk and financial risk tend to be tradeoffs in the risk behavior of farmers. Thus a decline in business risk tends to lead to the acceptance of greater financial risk thereby reducing the effects of the diminished business risk on the total risk.

In an attempt to verify the financial adjustment to changes in business risk, Gabriel and Baker

(1980) used linear regression to estimate the coefficients of the variables explaining changes in the component financial risk. The regression results indicated that in the aggregate, there was a financial response to changes in business risk: high (low) debt capital costs lead to high (low) financial risk ceteris paribus. In the aggregate, farmers make financial adjustments leading to decreased (increased) financial risk in response to a rise (fall) in business risk.

The Gabriel and Baker study dealt with aggregates and therefore net response of many decision makers. It is possible that certain categories of farmers do not respond as hypothesized. Furthermore the study focused more on the direction of the response than the magnitude of the response. Differential magnitudes in response may be associated with various characteristics such as risk aversion, farm size and farm type.

In making business decisions, net proceeds are usually calculated based on estimates and are therefore subject to a great deal of uncertainty. As a result, the decision to invest involves a greater or lesser degree of risk. Estimates of risk are dependent on the investor's attitude towards risk. If the market interest rate, i, is say 3%, for the investor to accept the proposal, the investor might require an expected rate of return on investment, ϒ, of say 5%, if the investor is very cautious, or 4% if less cautious. In other words, a risk premium is required, the size of which depends on what the risks are estimated to be and on the willingness to accept the risk.

MANAGEMENT

As a productive resource, management involves the organization and coordination of the productive resources by someone who makes the relevant decisions and carries them out. This fourth factor of production is concerned with the efficient allocation of the limited resources to achieve the stated objectives of the enterprise. Management is the risk-taker of the enterprise.

In economics, entrepreneurship has been traditionally categorized as one of the four factors of production. The National Knowledge Commission of India (NKC, 2008) defines entrepreneurship as the professional application of knowledge, skills, and competencies and/or of monetizing a new idea, by an individual or a set of people by launching an enterprise *de novo* or diversifying from an existing one (distinct from seeking self-employment as in a profession or trade), thus to pursue growth while generating wealth, employment and social good. If entrepreneurship is a latent form of behavior in all people just waiting to be evoked, then identifying the channels of directing the potential energy of entrepreneurship in order to achieve economic growth is a critical policy concern for both the public and private sectors (Stevenson, Roberts, Grousbeck, & Bhide, 1999). Bachenheimer, Isaak, and Isaak (2014), argue that the future wealth of particular regions in the world depends on each nation's human capital and this translates to the ability to cultivate and attract not only the brightest talent but also those with entrepreneurial capabilities.

Earliest definition of entrepreneurship refers to it as the process of bearing the risks of buying at certain prices and selling at uncertain prices. Distinguishing between the traits of the entrepreneurial personality, from managerial traits, Lynskey (2000) noted that while the latter is risk-averse and tends to be interested in the control of resources, the former searches for opportunities and so is a risk-taker.

Later definition of entrepreneurship refers to it as the creation of new enterprises and has been equated with starting one's own business. Schumpeter (1941), viewed entrepreneurship as a force of creative destruction. In this sense, established ways of doing businesses are destroyed by the creation of new and better ways of doing them.

There is a major difference between a small business with limited prospects for growth (where the business owner may have little or no desire for growth) and a new venture that is currently small. "Small business owner" tends to create a mindset of being small with limited scope for growth; an entrepreneur is by definition a risk-taker or someone who thinks out of the box (Abeysekera, 2011). Thus, in today's dynamic environment, one can conceive of a sustainable small business as a small entrepreneurial venture in which entrepreneurial behavior results in products or services of distinctive salience and scalability.

In the history of business, three basic forms of management have evolved: entrepreneurial management, traditional management, and small business management (Duening & Sheril, 2006). Entrepreneurial management focuses on growth; it entails organizing and managing business for profit by the entrepreneur who directly owns the means of production. Traditional management is focused on large company issues and challenges. In this case, assets (and of course liabilities) are organized and managed by professional managers whose job is to oversee production. It refers to the principles, techniques and approaches that are practiced by managers in large organizations. Small business management focuses on stability and maintaining a reliable customer base. A bit of overlap occurs in the three forms of management to the extent that they all deal with markets, customers, resources and strategies.

According to Duening and Sheril, (2006), the differences between entrepreneurial management and small business management are not obvious at the start-up stage. At the start-up stage, both are small and tend to grow rapidly. The degree of innovation tends to be an indicator of the type of management that is applicable. When small business becomes mature and profitable, the management's drive for growth tends to reduce dramatically. On the other hand, after the initial start-up growth stage had run its course, the entrepreneurial venture will continue to seek ways to grow rapidly.

Firms engaged in farm service marketing, agricultural production and food marketing are collectively known as agribusiness (Drummond & Goodwin, 2004). To all intents and purposes, this definition of agribusiness encompasses the activities of the small-scale farmers in agricultural production and food security. The small-scale farmers engage in the production of staple crops. In Nigeria, and especially in southeast Nigeria, one of the major crops small-scale farmers engage in is cassava (*Manihot spp*) production.

It is quite debatable whether or not the small-scale farming can be categorized as an entrepreneurial venture. In a related sense, it is quite debatable whether or not the small-scale farmer can be categorized as an entrepreneur.

The debate is similar to that about the difference between small business and entrepreneurship. Entrepreneurship implies three critical elements: new, sustainable and scalable. Whether all or some small-scale farmers can be viewed as an example of small entrepreneurial venture can be determined by the extent the small-scale farming business is initiating some aspects of the business that is new. This is not limited to a new product or technology; it includes exploring new and related opportunities. In terms of sustainability, we look out for a distinct competitive advantage; scalability presupposes the ability to achieve rapid growth over time.

It must be stated that all the four factors of production are not productive on their own. They work in combination with one another to achieve results. Acquiring the resources involves cost outlay. It makes good economic sense if the cost outlay in a typical farm firm is more than compensated by the profit made by it in given time period.

The drive to maximize profits tends to motivate the decisions in the farm firm. This is in line with the traditional theory of the firm. The two commonly canvassed criticisms the theory that profit maximization is a crude assumption about motivation and that the organizational structure of the firm affects its decision can be safely ignored. For one thing, profits undeniably define the life and the death of a farm firm. For another, the renewed emphasis on commercial agriculture presupposes that the farm firm should to all intents and purposes be market oriented and profit oriented. It is up to the entrepreneur to drive the enterprise towards the path of growth and expansion. This drive is encapsulated in the entrepreneurial spirit.

BRINGING ENTREPRENEURSHIP BEHAVIOUR TO AN ENTERPRISE

It is estimated that 70% to 80% of new businesses end their operations within their first year. Of those that survive, only half will remain in business after five years. In Nigeria, the price of oil, a principal and an important revenue earner, has been on a steady decline in the past couple of years. This decline has severely affected the provision of white collar jobs for the teeming labour force. Entrepreneurship is the best hope for reviving a weakened economy such as that of Nigeria. However, majority of those who initially start out as entrepreneurs rarely make it to the point of creating a sustainable competitive advantage and thereby demonstrate the ability to achieve rapid growth over time. In effect they tend to lack the entrepreneurial behavior and spirit for developing a distinct product or service and getting others to believe in it. This issue should therefore be addressed if the full potentials of entrepreneurship as they relate to reviving a weak economy are to be realized by the majority of those in Nigeria who set out to be entrepreneurs.

Basically, entrepreneurs care most about how to raise money, make money, save money and help others to do so. To accomplish this basic objective, conscious efforts must be made by the entrepreneur to bring entrepreneurship behavior to the business.

Entrepreneurship is the best hope for reviving a weakened economy such as that of Nigeria. Regrettably, majority of those who initially start out as entrepreneurs rarely make it to the point of creating a sustainable competitive advantage and thereby demonstrate the ability to achieve rapid growth over time. Regrettably, owners of most businesses tend to lack the entrepreneurial behavior and spirit for developing a distinct product or service and getting others to believe in the new product. It is the entrepreneurial drive that propels the business and eventually finds expression in enterprise expansion and growth.

If entrepreneurship is a latent form of behavior in all people just waiting to be properly channeled, then identifying the channels of directing the potential energy of entrepreneurship in order to achieve economic growth is a critical policy concern for both the public and private sectors. One of those channels is towards building and bringing entrepreneurship behaviour into an enterprise. It should be stated that entrepreneurial behaviour is not all there is for enterprise growth and expansion. It is however the mould in which all other requirements are cast and is therefore recommended for present and prospective entrepreneurs.

It must be realised that entrepreneurship is a multidimensional concept and its definition derives largely from the focus of the research undertaken (Verheul et al., 2001). For instance, Kaufman and Dant (1998) categorised contemporary definitions of entrepreneurship into three perspectives; traits, processes, and activities. There are demarcations such as the intellectual traditions of the German School (Schumpeter's innovation theory), the American School (Knight's uncertainty-bearing theory), and the Austrian School (Kirzner's market process theory). However, the exposition here is as the concept relates to small-scale agricultural enterprise that is expected to start small and grow big in line with its stated objectives.

Management Functions

As a decision-making process, management functions mainly involve coordination of the productive resources to achieve the desired objectives. These management functions can be encapsulated into three major decisions major decisions: investment decision, financing decision and dividend decision.

Investment Decision

This involves what to produce and allocating resources to achieve the best results. An aspect of the investment decision is capital investment. Capital investment is allocation of capital to investment proposals whose benefits are derived in the future. Investment decision also involves the decision to reallocate resources when an asset or a proposal is no longer economically viable.

Financing Decision

This involves the determination of the best financing mix or capital structure of the enterprise. While the finance structure of an enterprise refers to the different means used to raise capital, capital structure represents the proportionate relationship between the various long-term forms of financing. Any given combination of debt instrument and equity employed by an enterprise at any given point in time has important implications for solvency and profitability. The financing mix of the small-holder farmer tends to be made up of more of equity capital than debt capital.

Dividend Decision

The dividend decision includes the proportion of earnings paid to suppliers of equity, the stability of the earnings and the repurchase of stock if a quoted company.

IMPORTANCE OF FARM MANAGEMENT

As has earlier been stated, farm management focuses on the organization of land, labour, capital and human resources in operating a farm enterprise to meet the stated objectives. As a decision-making process, management functions mainly involve coordination of the productive resources to achieve the desired objectives. These management functions can be encapsulated into three major decisions major decisions: investment decision, financing decision and dividend decision. Against the foregoing background to the concept of farm management, the following considerations stand out as the major importance of farm management.

Driving Activities to Meet Set Business Objectives

The stated objectives of the farm have to be realized for the continued existence of farm firm. The realization of the set business objectives is largely dependent on the management making the right decisions and effectively executing the decisions. Farm management is therefore important to the extent that it drives the day-to-day activities of the enterprise towards the direction of meeting set business objectives.

Defining What Should Be Done for the Business Survival, Security and Expansion

Related to the directing activities towards meeting set business objective is the need to define what should be done for the business survival, the security of the business and enterprise expansion. Basically, the primary objective of an enterprise is to make profit. It should be stated that some firms are driven by the objective of maximizing its value to the shareholders; this value is in turn defined by the three major decisions that the firm must make namely investment, financing and dividend decisions. Related objectives being pursued by the average farm firm are those of maximization of food security and the stability of farm incomes. While profit maximization, food security and stable farm income objectives are not mutually exclusive, defining business goals and objectives also entail establishing what the business should do that will generate sufficient profit for enterprise expansion. Farm management is therefore important for defining what should be done for the business survival, security and enterprise expansion.

SOLUTIONS AND RECOMMENDATIONS

The financial service providers, in the exercise of their legitimate discretion and due diligence request clients to provide collaterals before funds are released. Banks require collaterals which the smallholder farmers find difficult to provide. It is recommended that such difficulty be avoided through group lending where members of a group of 5-10 come together to access a loan facility and thereby guarantee one another. It is for the sake of emphasis that this recommendation is highlighted here as it is made reference to in our discussions under Finance for Agriculture.

Present and prospective farm managers should be essentially have entrepreneurial spirit with which to start, grow and expand their enterprise in line with the objective of the enterprise in question.

FUTURE RESEARCH DIRECTIONS

The role of finance in enterprise expansion has always been emphasized. However, the fungibility of finance tends to be little understood. Future research is therefore recommended to be directed towards making farm managers to be aware of the inherent possibility of funds earmarked for a particular purpose can either by omission or commission be directed to another purpose with the attendant consequences of detracting from enterprise expansion.

CONCLUSION

Farm management defined as the organization of land, labour, capital and management, collectively called factors of production, in operating a farm enterprise to meet the stated objectives. Availability and efficient control and decisions over these resources go a long way in defining the attainment of the stated objectives of the farm enterprise. The four factors of production are not productive on their own as they work in synergy with one another for effectiveness. Acquiring the resources involves cost outlay. It makes good economic sense if the cost outlay in a typical farm firm is more than compensated by the profit made by it in given time period. As a decision-making process, management functions mainly involve coordination of the productive resources to achieve the desired objectives including enterprise expansion. Searching out investment opportunities and effectively ploughing back profits on existing or new investments for the expansion of the enterprise is one of the decisions of the management.

REFERENCES

Abeyekera, I. (2011). Reputation Building, Website Disclosure and the Case of Intellectual Capital. In *Studies in Managerial and Financial Accounting* (Vol. 21). Bingley, UK: Emerald Publishing.

Alimi, T. (2012). Small or Large Scale Agriculture for Nigeria: Issues, Challenges and Prospects. Inaugural Lecture Series 249. Obafemi Awolowo University Press.

Fabiyi, Y. L. (1979). Land Tenure Reforms in Nigeria: Implication of the Land Use Decree (Act) for Agricultural Development. *Ife Journal of Agriculture*, *1*(2), 235–257.

FAOSTAT. (2010). *Food and Agricultural Organization Statistical Year Book Food and Agriculture Organization, (FAO) (2014). Food and Agriculture: Key challenges and concrete solutions. The Food and Agriculture Organization of the United Nations (FAO) in collaboration with the Technical Centre for Agricultural and Rural Cooperation (CTA) and the International Fund for Agricultural Development* (IFAD).

Food and Agricultural Organization (FAO). (2012). *The state of food and agriculture*. Rome: FAO.

Gabriel, S. C., & Baker, C. B. (1980). Concepts of Business and Financial Risks. *American Journal of Agricultural Economics*, *62*(3), 110–152. doi:10.2307/1240215

Kaufmann, P. J., & Dant, R. P. (1998). Franchising and the Domain of Entrepreneurship Research. *Journal of Business Venturing*, *14*(1), 5–16. doi:10.1016/S0883-9026(97)00095-5

Oshio, P. E. (1990). The Indigenous Land Tenure and Nationalization of Land in Nigeria. *Boston College Third World Law Journal.* Retrieved from http://lawdigitalcommons.bc.edu/twlj/vol10/iss1/3

Upton, M. (1973). *Farm Management in Africa the Principles of Production and Planning*. Oxford University Press London.

Verheul, I., Wennekers, S., Audretsch, D., & Thurik, R. (2001). *An Eclectic Theory of Entrepreneurship: Policies, Institutions and Culture. Research Report 0012*. Zoetermeer: EIM.

Chapter 4
Financing and Training Imperatives for Resilient Agriculture in Nigeria

ABSTRACT

This chapter focuses on the financing and training imperatives for resilient agriculture in Nigeria. A proposed twin-track approach to addressing the challenge of agriculture as a development issue in Nigeria involves both encouraging agri-business and supporting the large population of smallholders. On the basis of supporting the large population of smallholder farmers, this chapter highlights the financial and training needs for the efficient practice of resilient agriculture. Discussions in this chapter are based on secondary data obtained from relevant sources. Supporting farmers for the purpose of resilient agriculture requires conscious investment in generation, development, and dissemination of relevant agricultural practices for agricultural intensification and sustainable use of natural resources. In effect, finance and training in support of the farmers are called for. Against the background that there is low government investment in agriculture, and because agriculture is a private sector activity, special funding mechanisms that incorporate public-private partnerships that deliver financial solutions are recommended. With respect to training, resilient agricultural practices should form part of the curriculum in the primary, secondary, and tertiary schools in Nigeria.

DOI: 10.4018/978-1-5225-3059-6.ch004

INTRODUCTION

In most developing countries including Nigeria, the agriculture sector's role goes beyond the classical functions associated with providing food, raw materials, employment and incomes. According to Eboh (2011), the roles of agriculture in a modernizing economy are much more transformative than are implied in the classical models of development. The topical transformative role of the agricultural sector is typified by the fact that it is at the centre of the strategy for economic diversification from oil.

As the population increases, Nigeria will be compelled to find ways and means to effectively transform the agriculture sector to adequately meet the needs of this population while simultaneously decreasing the environmental impact attendant to the transformation programme. Additionally, climate change clearly compounds the challenges confronting the agriculture sector. It has been argued that to assure food security, eradicate rural poverty and create social stability, policies and institutions are required to enhance the ability of individuals, households and production systems to recover from the incidence of shocks and stresses on the agriculture sector as a result of climate change.

Globally, there is high pressure on land use as 20% of forests and 50% savannahs, grasslands have been converted to agriculture (Ramankutty & Rhemtulla, 2012; Pereira, 2010). Land cover change for agriculture, which is one of the key drivers of biodiversity loss, could increase extinction rates a hundred-fold over the 21st century (Baumert, Herzog, and Pershing, 2005). Agriculture is responsible for more greenhouse gas emissions than any other human activity (Baumert, Herzog, and Pershing, 2005). In Nigeria, agriculture contributes a small extent to global warming through bush burning and other poor land management practices, however it bears the full brunt of climate change impacts. For instance, according to Federal Ministry of Agriculture and Rural Development (FMARD) (2014), the 2012 flood, in which over 2.3 million people were displaced, 363 persons lost their lives and the total value of losses across all sectors of the national economy put at US$16.9 billion, exposed Nigeria's vulnerability to extreme climate events and underscores the need for deliberate planning to enhance adaptive capacity and resilience across all sectors of the Nigerian economy, including agriculture. Without doubt, agriculture is critical and central for Nigeria's overall development; it is also a major driver of environmental decline.

An approach aimed at developing technical, policy and investment conditions to achieve sustainable agricultural development for food security under climate change is the notion of climate smart agriculture. Sponsored by the Food and Agricultural Organization (FAO) climate smart agriculture entails (i) sustainably increasing agricultural productivity and incomes; (ii) adapting and building resilience to climate change; and (iii) reducing and/or removing greenhouse gases emissions, where possible (FMARD, 2016). To all intents and purposes, resilient agriculture is an integral part and parcel of climate smart agriculture.

According to Nwajiuba (2012), 'Nigeria needs a programme devoted to a new generation of farmers that trains young educated people interested in agricultural entrepreneurship and provides financial and technological support. They should be enabled to use improved technologies and modern management approaches that help ensure farm profitability and sustainable resource use'. The foregoing views were stated in the context of a proposed twin-track approach to addressing the challenge of agriculture as a development issue in Nigeria. The twin-track approach involves both encouraging agri-business and supporting the large population of smallholders which according to Nwajiuba (2012), is critical to rural food security, social cohesion and poverty alleviation. On the basis of encouraging agri-business and supporting the large population of smallholder farmers, this paper seeks to highlight the financial and training support needs for the efficient practice of resilient agriculture.

The rest of the paper is organized as follows: what resilient agriculture is, enabling environment for agricultural resilience, financing and training needs for resilient agriculture. The paper concludes with recommendations.

RESILIENT AGRICULTURE

Resilience can be viewed as the presence or absence of such capacities as: capacity to react to a phenomenon or a disturbance and prevent or lessen harm to an existing state; the capacity to effortlessly rebound to a stable state after a disturbance to an existing system; and the capacity to alter an existing state to one that is more responsive to disturbance. The presence of these capacities can be described as resilience while absence can be referred to as not being resilient. According to Vallee (2008) resilience is the capacity of a system to experience change while retaining essentially the same function, structure, feedbacks, and therefore identity. It is the ability of a system to withstand or absorb stress or shocks and adapt to change.

Resilient agriculture refers to the capacity of agricultural development to resist or recover from stresses and shocks and so rebound to the previous level of growth. According to Bennett et.al., (2014), a resilient agriculture is agriculture that meets the food and the development needs of the people over both the short-term and the very long-term, from local, though, state, national, regional, global scales, without destabilizing the earth system. Resilient agriculture aims to maintain or grow the full natural capital of the ecosystem. Additionally, it also aims to maintain and grow such broader set of mechanisms as social networks, governance, and leadership required to meet the immediate needs of the people without undermining the long-term stability of social and natural systems that together provide services to humanity. Resilient agriculture creates agricultural growth out of knowledge, investment and innovation, while simultaneously building the capacity of farmers, particularly smallholder farmers to counter environmental degradation and climate change (Agriculture for Impact, 2012).

Bennett et al. (2014), stated that achieving the twin goal of increasing agricultural production and reducing its environmental impact is based on increasing the efficiency of agricultural production relative to resource use and relative to unintended outcomes such as water pollution, biodiversity loss, and greenhouse gas emissions. Bennett et al. (2014), further stated that while increasing agricultural production efficiency is certainly necessary, it is unlikely to be sufficient and may in some instances reduce long-term agricultural resilience.

According to Agriculture for Impact (2012), the goal of resilient agriculture may seem at odds with agricultural growth. It is possible to experience a highly resilient agriculture but stagnant agricultural growth, or a rapid agricultural growth that is destructive and highly volatile. The ideal is somewhere in between where appropriate resilience is built into growth at the outset in a way which exploits the synergies between growth and resilience.

Modern agriculture has substantial impacts on the biophysical environment. These impacts can work against the very processes that support the functioning of agricultural systems, and consequently reduce the long-term sustainability of agriculture itself (Zhang *et al.*, 2008). Therefore, there should be conscious efforts toward sustainable intensification of agriculture which aims to optimize crop production per unit area while accounting for social, political, and environmental impacts. To all intents and purposes, these conscious efforts should be in the direction of policies and practices that focus on increased production efficiency at lower environmental and resource costs (Godfray & Garnett, 2014).

Resilient agriculture essentially entails our conscious efforts to build an agricultural production system capable of meeting current, forthcoming, known and unknown challenges. Resilient agriculture requires that we modify our views on sustainable agricultural development from its current emphasis on optimization of production in relation to immediate economic, social, and environmental costs. For agriculture to be actually profitable it must be resilient; for it to be resilient, cost considerations must transcend immediate economic costs and necessarily focus on environmental cost. Successful agriculture is most often perceived as high yields with low economic cost. Increasingly, this also includes high yields at low environmental cost.

To round off the discussion on resilient agriculture, it is pertinent the highlight differences between the concept of sustainable agriculture and the concept of resilient agriculture. Under the Green Revolution, appreciable increases in agricultural productivity were achieved in the 20th century. The increases were achieved through the adoption of adoption of such practices as mono-cropping and use of pesticides and fertilizers. The unintended costs associated with the yield increases included degraded soil qualities, biodiversity loss and diminished pest resistance. According to IFAD (2012), increased pesticide and fertilizer pollution in soils and ground water put both the environment and human health at risk.

Climate-smart agriculture (CSA) is an improvement on the concept of Green Revolution and has some features in common with sustainable agriculture approaches. CSA is built on such technical foundation as sustainable agriculture, sustainable intensification and conservation agriculture (CSA Guide, 2017).The three essential features that differentiate CSA from sustainable agriculture are: 1. An explicit focus on climate change, 2. Search for synergies and negotiations of trade-offs focusing on productivity, adaptation and mitigation outcomes and 3. The availability of new funding opportunities for agricultural development. Let us take a close look at these three features which differentiate CSA from sustainable agriculture.

Pursuit of Climate Change

In line with other sustainable agricultural approaches, CSA is embedded on the principles of increased productivity and sustainability. It is marked out by emphasis on climate change, clearly addressing adaptation and mitigation challenges targeted at food security. The extent to which CSA is related to sustainable agriculture, incorporates resilience concerns and at the same time seeks to reduce greenhouse gas emissions can be presented in an explicit equation (CSA Guide, 2017).

CSA = Sustainable Agriculture + Resilience – Emissions.

The above equation means that CSA is equal to sustainable agriculture which encompasses resilience features and reduced greenhouse gas emissions.

Synergies and Trade-Offs With Respect to Productivity, Adaptation and Mitigation Outcomes

CSA essentially focuses on the development of technologies and practices that simultaneously meet the three challenges of productivity, adaptations and mitigation at the farm level in addition to the outcomes of the interventions beyond the farm. To all intents and purposes, CSA considers the synergies and trade-offs that exist among productivity, adaptation and mitigation, in addition to the interactions that occur at the wider socio-ecological levels. Farmers and decision-makers should therefore understand the synergies and trade-offs that exist among the three pillars and within different levels.

Available Funding Opportunities

On account of the emphasis on climate change, CSA opens up new funding opportunities for agricultural development, by allowing the sector to tap into climate finance for adaptation and mitigation (CSA Guide,2017). Examples of the funding sources include: the Adaptation Fund, the Least Developed Countries Fund or the Special Climate Fund, as well as the Clean Development Mechanism and the Voluntary Carbon Market. Most of all is the earmarked allocation which has been made specifically for CSA by the Global Environment Facility Trust Fund (GEF) and the future Green Climate Fund (CSA Guide, 2017).

ENABLING ENVIRONMENT FOR RESILIENT AGRICULTURE

To achieve resilient and sustainable agriculture, we should search beyond a one-size fits all single solutions, and instead move towards a mosaic of solutions operating across scales. The Montpellier Panel report on how to achieve agricultural growth with resilience recommends that governments working in partnership with local and international private sector organizations should aim at building resilient markets, resilient agriculture and resilient people (Agriculture for Impact, 2012). Resilient market entails reduction in price volatility, facilitating private sector investment in agriculture, and building better environments by governments and private sectors. Enabling environment can be construed against the background that the operation of market forces alone, whether within a country or on a global scale, will not create food security; benefits will only flow if there is an appropriate enabling environment. Resilient agriculture entails enabling resilient and sustainable intensification, combating land and water degradation and building climate smart agriculture. Having resilient people encompasses scaling up nutrition, focusing on rural women, and youth, and building diverse livelihoods. With its focus on the markets, the agricultural system and the people, Agriculture for Impact (2012), clearly brings out the need for integrated agricultural intervention plan that will bring about agricultural growth with resilience.

Governments at all levels working in active partnership with local and international private sector organizations require significant investments to bring about resilient markets. Similarly, appreciable investments are required for the generation, dissemination and adoption of resilient agricultural technologies and practices build on agro-ecological knowledge. Making people resilient requires reasonable investment in capacity building and income diversification. In all, a common thread runs the recommendations for the achievement of agricultural growth with resilience: finance for investments in human and natural resources. Investment in human resources also entails capacity building and training.

FINANCING AND TRAINING IN SUPPORT OF RESILIENT AGRICULTURE

Let us first look at the fundamentals for financing resilient agriculture in Nigeria.

Fundamentals for Financing Resilient Agriculture

The basis for government investment in the agricultural sector is rooted in the fundamental justifications for public-sector intervention in the macro economy. According to Olomola *et al.* (2014) arguments in support for public-sector involvement in the economy include the market failure argument and the inequality or poverty argument. The market failure as a concept clearly comes about when the competitive market system either produces the "wrong" amounts of goods and services or fails to allocate any resources whatsoever to the production of some economically justified goods and services (McConnel & Brue, 1999). Oji (2002), identified a third type of market failure arising from imperfect information available to producers and consumers of goods and services and concluded that governmental action or policy can correct all types of market failures.

The inequality or poverty argument is founded on the basis that majority of the population in developing countries are small-holder farmers who are engaged in agriculture. Therefore, incentivizing and subsidizing the practice of agriculture improves incomes and strengthens livelihoods of the majority of the people thereby alleviating poverty should essentially concern a responsible government. Small-holder farmers usually face various forms of poverty such as weak property rights, poor access to markets and financial services, and severely limited ability to manage risks (FAO, 2012). It has been argued that low income leads to low savings, low savings give rise to low investment, low investment translates to low productivity, low productivity engenders low income and so the cycle continues. This vicious circle of poverty can be broken by the external injection of production capital which will expectedly and ultimately increase productivity, increase income, increase savings and investment and lift the entity concerned out of the poverty level. In agricultural production process, this external injection of capital comes in the form of agricultural credit. It therefore stands to reason that agricultural credit, when properly administered, plays a role in breaking the vicious circle of poverty.

Federal government spending on agriculture is low in Nigeria. Olomola *et al.* (2014) in a study observed that between 2008 and 2012, the federal agriculture budget and actual expenditure shares of total expenditure were, on the average, 5.5 and 4.6 percent, respectively. At the state level, the trend is not different. In Cross River state for instance, according to Olomola *et al.* (2014), the agricultural sector budget, on the average, accounted for 1.2 percent of the total state budget between 2008 and 2012. The share of actual agricultural expenditure in actual total state expenditure was 0.9 percent in the same period. Similarly, agricultural spending as a share of total at the Local Government Areas unfavorably compares with that of other sectors. In Akamkpa LGA, for instance, agriculture received the least attention in terms of actual expenditures (1 percent on average between 2008 and 2012), while most of the funds (29.1 percent on average during the same period) were expended on works. Local government has no capacity to formulate policies; neither does it have the financial independence to meaningfully execute an agricultural budget. Olomola *et al.* (2014) concluded that despite the importance of the agricultural sector in the Nigerian economy and the priority often accorded it in development strategies, the sector remains grossly underfunded.

Budgetary allocations to the agricultural sector be they high or low, do not on their own equate to financing climate resilient agriculture. While climate resilient agriculture focuses on the interface between climate change and agriculture, funds for agricultural development, climate mitigation and climate adaptation generally come from different sources. Obviously, the low budgetary allocation to the agricultural sector is a reflection of the funding gap in financing climate resilient agriculture. For instance, ACARN (2014), identified the absence of nationally generated figures for the funding climate change related programs, as they relate to both mitigation and adaptation in Nigeria. According to ACARN (2014), out of the global total flows of climate finance of approximately USD 364 Billion in 2011, close to about 26% came from the public sector (domestic, bilateral and multilateral) while 74% came from private sector finance.

The low budgetary allocation to agriculture is both a challenge in financing the required investments for the achievement of agricultural growth with resilience and also a clear indication for the need to additionally explore innovative financing mechanisms. The innovative financing mechanisms are in no way a substitute or an alternative to public funding. Innovative financing mechanisms are innovations to the extent that they depart from traditional approaches to mobilizing development finance.

According to Girishankar (2009) innovative financing entails non-traditional applications of solidarity, public private partnerships, and catalytic mechanisms that (i) support fund-raising by tapping new sources and engaging investors beyond the financial dimension of transactions, as partners and stakeholders in development; or (ii) deliver financial solutions to development problems on the ground. According to Sandor, Scott, and Benn (2009), examples of innovative financing include new approaches designed to attract private and public revenues for funding activities that benefit partner countries; new revenue streams (e.g. voluntary contribution scheme, a new tax etc) devoted to developmental activities on a multi-year arrangement; and new incentives (corporate social responsibility, rewards etc) to address market failures .

Innovative financing is essential especially to the extent that it enables the mobilization of private funding. According to Agriculture for Impact (2012), by its nature agriculture is a private sector activity. This is so because almost all the organizations that produce agricultural goods and services are owned and operated by private entities and units such as firms and households. The private sector in Nigeria therefore has a lot to offer by way of investments in agriculture.

Fundamentals for Training for Resilient Agriculture

According to FAO (2014), there is the challenge of youth insufficient access to knowledge, information, education and training. Poor and inadequate education and training limit productivity and acquisition of skills. To raise a new generation of young educated farmers that are interested in the use of improved technologies and modern management approaches that help ensure farm profitability and sustainable resource requires training. It also requires re-orientation. Building resilient people who will practice resilient agriculture embraces building the capacities of the people. Training and education are needed to develop new and relevant technologies and processes. Training and education are also needed to efficiently disseminate and effectively adopt the new technologies and processes.

According to agriculture for impact (2012) resilient people are capable of generating profitable and diverse livelihoods that ensure stable incomes, sufficient nutrition and good health in the face of recurrent stresses and shocks. Against the backdrop of the role of agriculture in Nigeria, there is an obvious need to improve present and prospective farmers' (especially women and youth) access to the requisite skills and competences relating resilient agricultural practices. Agriculture for Impact (2012), specifically recommended that governments in partnership with the private sector and NGOs should enable access to education and training for women and youths from the outset in all development programmes.

In addition to training and re-training of present and prospective farmers, a critical and rigorous re-assessment of the quality and content of teaching and learning of agriculture in all levels of education in Nigeria is also fundamental to the achievement of resilient agriculture in Nigeria. According to International Chamber of Commerce (ICC) (2011), education is paramount in making green economy operational– green economy, a related concept to resilient agriculture, defined as an economy that improves human well-being and social equity and significantly reducing environmental risks and ecological scarcities.

SOLUTIONS AND RECOMMENDATIONS

Based specifically on the financing and training needs in support of the practice of resilient agriculture, the following recommendations are offered.

- There is the need for Nigeria to design special funding mechanism that is suited for both finance food security and climate resilient agriculture. The special funding mechanism can be established by exploring such innovative finance options as sovereign funds, specific forms of taxation etc. the special funding mechanism should not just be financing sources but effectively provide financial solutions to the issues of climate resilient agriculture.
- Private sector investment in financing agriculture is recommended. The private sector can provide funding and invest in the production of relevant and appropriate technologies that both increase production and make agriculture more resilient and sustainable.
- In the long term and with respect to training, ensuring resilient agriculture will necessitate a renewed focus on the quality and content of instructions in institutions of learning as they relate to resilient agriculture. The learning experiences that are provided in our institutions should open opportunities for learning and for new practices to emerge. Efforts towards resilient agriculture to all intents and purposes involve basically new approaches. We should lay the foundation for these new approaches through teaching and learning in and out of school. Out of school agricultural extension services should, in both the short and the long terms, be sufficiently enriched with resilient agricultural practices. Resilient agricultural practices should form part of the curriculum in the primary, secondary and tertiary schools in Nigeria.
- In all, the role of government in financing agriculture and in providing the enabling environment cannot be overemphasized.

FUTURE RESEARCH DIRECTIONS

Although agriculture is a private sector activity, it is necessary that the government should create the enabling environment. In creating the enabling environment, relevant policies are created for implementation. However, these policies are dropped as soon as the government that created them is no longer in power thus giving rise to policy reversal. There is the need to investigate into the incidence of policy reversal and how to stem the trend.

CONCLUSION

Resilient agriculture entails enabling resilient and sustainable intensification, combating land and water degradation and building climate smart agriculture. To achieve resilient agriculture, special funding mechanism and formal and informal agricultural training programmes for present and prospective farmers especially women and youth are advocated.

REFERENCES

Abiodun, B. J., Lawal, K. A., Salami, A. T., & Abatan, A. A. (2013). Potential influences of global warming on future climate and extreme events in Nigeria. *Regional Environmental Change*, *13*(3), 477–491. doi:10.1007/s10113-012-0381-7

Abiodun, B. J., Salami, A. T., Mathew, O. J., & Odedokun, O. D. (2012). Potential impacts of afforestation on climate change and extreme events in Nigeria. *Climate Dynamics*. doi:10.1007/s00382-012-1523-9

Advisory Committee on Agricultural Resilience in Nigeria (ACARN). (2014). *National agricultural resilience framework (NARF)*. Abuja, Nigeria: Federal Ministry of Agriculture and Rural Development.

Agriculture for Impact. (2012). *Growth with resilience: Opportunities in African agriculture*. The Montpellier Panel London.

Agwu, N. M., Nwankwo, E. E., & Anyanwu, C. I. (2014). Determinants of agricultural labour participation among youths in Abia State, Nigeria. *International Journal of Food and Agricultural Economics*, *2*(1), 157–164.

Baumert, K. A., Herzog, T., & Pershing, J. (2005). *Navigating the numbers: Greenhouse gas data and international climate policy*. Washington, DC: World Resources Institute.

Bennett, E., Carpenter, S., Gordon, L., Ramankutty, N., Balvanera, P., Campbell, B., . . . Spierenburg, M. (2014). Toward a more resilient agriculture. *Solutions, 5*(5), 65-75. Retrieved from http://www.thesolutionsjournal.com/node/237202

Central Bank of Nigeria (CBN). (2011). *Annual report and statement of accounts*. CBN Abuja.

Cervigni, R., Rogers, J. A., & Henrion, M. (2013a). *Nigeria: Toward climate-resilient development in Nigeria*. Washington, DC: World Bank.

Cervigni, R., Rogers, J. A., & Henrion, M. (2013b). *Nigeria: Opportunities for low-carbon development*. Washington, DC: World Bank. doi:10.1596/978-0-8213-9925-5

CSA Guide. (2017). *How is climate-smart agriculture different from other sustainable agriculture approaches?* Available at: https://csa.guide/csa/how-is-it-different.

Eboh, E. C. (2011). *Agricultural economy of Nigeria: Paradoxes and crossroads of multimodal nature. 56th Inaugural Lecture of the University of Nigeria*. University of Nigeria Press Limited.

Federal Ministry of Agriculture and Rural Development (FMARD). (2014). *Agricultural Resilience Framework*. Federal Ministry of Agriculture and Rural Development.

Federal Ministry of Agriculture and Rural Development (FMARD). (2016). *The Agriculture promotion policy (2016 – 2020) building on the successes of the ATA, Closing key gaps policy and strategy document*. Federal Ministry of Agriculture and Rural Development.

Food and Agricultural Organization (FAO). (2012). *The state of food and agriculture*. Rome: FAO.

Food and Agricultural Organization (FAO). (2014). *Youth and agriculture: key challenges and concrete solutions*. Rome: FAO.

Girishankar, N. (2009). *Innovating development finance from financing sources to financial solutions* World Bank.

Godfray, H. C. J., & Garnett, T. (2014). Food security and sustainable intensification. *Philosophical Transactions of the Royal Society of London. Series B, Biological Sciences, 369*(1639), 20120273–20120273. doi:10.1098/rstb.2012.0273 PMID:24535385

International Fund for Agricultural Development (IFAD). (2011). *Innovations in financing food Security*. Rome: IFAD.

International Fund for Agricultural Development (IFAD). (2012). *Climate-smart smallholder Agriculture: What's different?* IFAD Occasional Paper. Rome: IFAD.

McConnel, C. R., & Breu, S. L. (1999). Macroeconomics: Principles, Problems and Polices (14th ed.). McGraw Hill Coy Inc.

Nwajiuba, C. (2011). *Nigeria's agriculture and food security challenges*. Retrieved from http://www.Chapter_--_4_Green_Deal_Nigeria_AGRICULTURE.pdf

Oji, K. O. (2002). *Basic Principles of Economics for Agricultural Projects and Policy Analyses*. Nsukka, Nigeria: Prize Publishers.

Olomola, A., Mogues, T., Olofinbiyi, T., Nwoko, C., Udoh, E., Onu, J., & Woldeyhannes, S. (2014). *Analysis of agricultural public expenditure in Nigeria examination at the federal, state, and local government levels*. IFPRI Discussion Paper 01395. The International Food Policy Research Institute (IFPRI).

Pereira, H. M., Leadley, P. W., Proença, V., Alkemade, R., Scharlemann, J. P. W., Fernandez-Manjarrés, J. F., & Walpole, M. et al. (2010). 2 Scenarios for global biodiversity in the 21st century. *Science, 330*(6010), 1496–1501. doi:10.1126/science.1196624 PMID:20978282

Ramankutty, N., & Rhemtulla, J. M. (2012). Can intensive farming save nature? *Frontiers in Ecology and the Environment, 10*(9), 455–455. doi:10.1890/1540-9295-10.9.455

Sandor, E., Scott, S., & Benn, J. (2009). *Innovative Financing To Fund Development: Progress and prospects*. Retrieved from www.oecd.org/dac

Vallée, J-C. (2008). *Resilient agricultural systems*. Retrieved from www.cielap.org/pdf.Resilience_LeVallee.pdf

Zhang, W., Ricketts, T. H., Kremen, C., Carney, K., & Swinton, S. M. (2007). Ecosystem services and dis-services to agriculture. *Ecological Economics*, *64*(2), 253–260. doi:10.1016/j.ecolecon.2007.02.024

Chapter 5
Enterprise Expansion and Opportunities for Expansion in Agriculture

ABSTRACT

It has often been stated that the agricultural sector has the potential to provide the needed raw material for the manufacturing sector. It is pertinent to clearly identify this potential, interrogate why it still remains potential, and more importantly, suggest workable ways to sustainably and profitably exploit the potential as a going concern. This chapter is therefore designed to focus on enterprise expansion and opportunities for expansion in agriculture. The chapter is divided into the following sections: "Characteristics of Agriculture in Nigeria," "Determinants of Youth Participation in Agriculture," "Agricultural Enterprise Expansion and Agricultural Transformation," "Enhancing Enterprise Expansion," and "Opportunities in Agriculture." The chapter concludes that for the potential of agriculture in an economy to be realized, the relevant stakeholders should know that business as usual is not an acceptable option; yield-increasing and enterprise-expansion-inducing strategies should be implemented in both the short and the long term. Recommendations are made to enable those engaged in agriculture to profit by it and increase in both output and in scale.

DOI: 10.4018/978-1-5225-3059-6.ch005

INTRODUCTION

This chapter focuses on engaging in agriculture and successfully growing it as a business. It looks at what should be emphasized in our current drive for expanding agriculture. The chapter also focuses on the opportunities available in agriculture which could be latched for expansion with a business sense. The rest of the chapter is organized as follows: Characteristics of agriculture in Nigeria, determinants of youth participation in agriculture, agricultural enterprise expansion and agricultural transformation, enhancing enterprise expansion, opportunities in agriculture, solutions and recommendations, future research directions, and conclusion.

BACKGROUND

Agriculture continues to be the largest employer in Africa, but accounted for only 13 percent of GDP in 2010 (Fox et al. 2013). In 2016, agriculture's contribution to the GDP in Nigeria was 24.43% (NBS, 2017). Brooks et al. (2013a) argue that accelerated transformative change in the agricultural sector will lead to an increase in productive capacity, reduction in real food prices, and rural income inequality. The transformative change is also expected to engender agricultural job creation. According to Adesugba and Mavrotas (2017), job creation should not just be targeted towards the wage job sector, but should also seek to increase job opportunities in agriculture in both the formal and informal sub-sectors.

The agricultural sector has the potentials to provide the raw materials needed for the manufacturing sector. It should be noted that the potential to create jobs in the agricultural sector is increased in countries that have productive farms with better infrastructure and communication networks. It is pertinent to clearly identify these potentials, interrogate why they still remain potentials and more importantly suggest workable ways to sustainably and profitably exploit the potentials as a growing concern. Let us take a look at the characteristics of agriculture in Nigeria as a point of departure on the discussion on enterprise expansion and opportunities in the sector.

Characteristics of Agriculture in Nigeria

Nigerian agriculture can be described by the following characteristics:

Agricultural production activities especially crop production are rain fed. The practice of agriculture in Nigeria is solely dependent on the availability and even the unavailability of natural rainfall. The irrigation facilities where they exist especially in the northern parts of the country are in various stages of obsolescence. In the southeastern Nigeria there is no irrigation facility and to that extent both wet and dry seasons farming are dependent on the vagaries of the rain.

Agriculture is also characterized by low input use and productivity is generally low. In many areas agricultural practices are on extensive and inefficient basis with less than satisfactory use of fertilizer. The low input use and concomitant low productivity is a graphical illustration of the vicious cycle as low input use give rise to low output.

Furthermore, technology applied in Nigerian agriculture is simple, rudimentary with little or no mechanization. Lack of mechanization is closely related to the size of per capita land holdings and the scattered and fragmented nature of the holdings. It is a reflection of the attention paid to the development of the sector by past and present governments in Nigeria. The implication is that almost all the agricultural practices are undertaken by manual labour with the obvious limited efficiencies that go with such practices. In early classical economic theory, agriculture was characterized as a sector with low productivity, traditional technology and decreasing returns (Nin-Pratt, 2010). The simple and rudimentary nature of the technology applied in Nigerian agriculture is not appreciably far from that described by the early classical economic theory.

Another obvious feature of Nigerian agriculture is the predominance of subsistence agriculture with little or no market orientation. There is therefore the need for conscious and workable strategy by the relevant stakeholders to aid the transition from subsistence farming to small, medium and large scale commercial agriculture. According to FMARD (2010), the transition can be facilitated by helping deal with market failures that continue to limit smallholder participation in the market. The emphasis on market is expected to balance the relative role of supply and demand factors in enhancing growth of the agriculture sector

A somewhat specific feature of land ownership in Nigeria is the fragmentation of farm plot into uneconomic units that are not amenable to mechanization. According to NARP (1994), more than 70 percent of the farming population in Nigeria is made of smallholder farmers; each owns or cultivates less than 5 ha of farmland. However, the smallholders account for 90 percent of the total farm output (FMARD, 2010).

Another characteristic is the low investment below the pledged 10% of the budget. In 2013 Nigeria's budgetary allocation to agriculture budget was 2%. In the 2016 Appropriation Act, out of a total of N 6,060,677,358,227 being the consolidated revenue fund of the Federation for the year, the sum of N 29,632,584,416 and N46,173,963,859 only were allocated to the Federal Ministry of Agriculture as recurrent and capital expenditures respectively (Appropriation Act 2016, n.d,) The total budgetary allocation to the Federal Ministry of Agriculture in 2016 was N75,806,548,275, equivalent to 1.25% of the national budget.

At the Second Ordinary Assembly of the African Union in July 2003 in Maputo, African Heads of State and Government endorsed the "Maputo Declaration on Agriculture and Food Security in Africa" (Assembly/AU/ Decl. 7(II). The Declaration contained several important decisions regarding agriculture; prominent among them was the "commitment to the allocation of at least 10 percent of national budgetary resources to agriculture and rural development policy implementation within five years" (NEPAD, 2003). At the commencement of the 23rd Ordinary Summit of the African Union (AU) in Malabo Equatorial Guinea the terms of the 2003 Maputo Declaration of Food Security and Agriculture in Africa were reaffirmed in a commitment deemed the 2014 Malabo Declaration. The implication is that the declaration is as important today as it was in 2003. Investment in agricultural capital is therefore a prosperous answer to the question of transforming and expanding agricultural production and practices.

Another worrisome feature is that youth are leaving agriculture in Nigeria. In Nigeria, there are clear signs that the exit from Southern Nigeria's youth is more pronounced than the exit from their Northern Nigeria counterparts (Agwu et al, 2014). The exodus of youth from agriculture invariably creates a generation of older farmers who tend to be technologically less savvy than the younger farmers

Determinants of Youth Participation in Agriculture

Several reasons may explain why the youth would be engaging less in agriculture in Nigeria. Structural transformation theory suggests people switch out of low productivity sectors when structural transformation occurs. As a country's incomes rise, labor reallocation takes place on account of productivity gaps between sectors. In developing countries, agriculture is generally the sector with the lowest productivity. So, as a country develops, productivity differences between agriculture sector and the non-agricultural sector in an economy tend to lead to better wages in the latter and labor will progressively move out of agriculture. So one reason youth do not take up agricultural practices would be structural transformation. The structural transformation theory may not be applicable to the Nigerian situation because the agricultural sector is neglected as opposed to have been transformed to release labour to other sectors of the economy.

Population pressure leads to land scarcity issues (Pingali et al., 1987). To make a livelihood out of agriculture one needs to acquire or rent land. However, with ill-defined land tenure policies in many African countries, land markets are very incomplete hindering the ability of individuals to buy or rent land (Amanor, 2010). A recent finding that supports the land access hypothesis is the one by Bezu and Holden (2014) who concluded based on econometric analysis that the main reason for rural Ethiopian youth leaving agriculture is the lack of access to land.

Sumberg et al. (2014) argue that interventions targeted at providing jobs for youth by trying to increase their involvement in agriculture operate on the basis that educated youth do not find agriculture attractive. Such interventions are then designed to try to deal with the image problem of agriculture.

Agwu et al. (2014) investigate the determinants of youth participation in agricultural labour in Abia State, Nigeria. Using a probit model, they found that own education, non-agricultural income, parents' occupation, father's education, farm size and mechanization rate significantly impact the likelihood of youth participation in the agricultural labor force in Abia State.

The characteristics necessary for the development of the agricultural sector in Africa are also important for absorbing youth into the sector (Brooks et al., 2013b). This is also applicable to the agricultural sector in Nigeria. According to Adesugba and Mavrotas (2017), agriculture has to be able to compete with other sectors in a profitable manner in order to engender accelerated transformative change to improve productivity and rural incomes to create jobs for youth. Evidently, policies and programmes that increase the participation of youth in agriculture will enhance the job creation performance of the sector.

The characteristics of agriculture in Nigeria if not radically and consciously altered cannot engender the much-needed growth and expansion. For agriculture to be a profitable venture, a going concern, will ever remain just potentials. It should be noted that the transformative approaches should be affected along the lines of the identified obviously limiting characteristics. Let us look at some issues that are pertinent to agricultural expansion and the related term of agricultural transformation.

Agricultural Enterprise Expansion and Agricultural Transformation

Agricultural enterprise expansion is growth of an agricultural business' product and service offerings in a given location. It can be measured in terms of sustained increases in the quantity and quality of goods and services produced by a specific agricultural enterprise. Agricultural expansion can also be exemplified by the opening up of additional farms, stores, outlets as the case may be as dictated by the demands of market and in line with the overall goal of the enterprise. Agricultural enterprise expansion is a stock concept in which growth is achieved as exemplified in increases in scales. Agricultural enterprises can experience expansion, contraction or even stagnation. In line with the focus of this book, we are looking at the enabling environment for agricultural enterprise expansion.

For agricultural products, there is a limit to which growth of an agricultural business' product and service offerings in a given location ultimately enhances the profitability of the farmer. For agricultural products that are highly perishable, it makes no economic sense for a farmer to experience increases in output when there is no guaranteed ready market. Where there is growth of an agricultural business' product and service without the desired value chain support, the output will be sold at give away prices. At worst they output will be sold at no prices in which they are left to rotten and go bad.

An agricultural value chain can be described as a set of interlinked activities that operate to add value to a product. Agricultural value chains encompass the flow of products, knowledge and information between smallholder farmers and consumers (http://ag4impact.org/sid/socio-economic-intensification/building-social-capital/agricultural-value-chains). Agricultural value chain is made up of producers, processors and processes that effectively and efficiently improve a product and ultimately linking commodity producers to processors and markets. According to Conway (2012), agricultural value chains offer the opportunity to capture added value at each stage of the production, marketing and consumption process. Farmers especially smallholder farmers need to better engage with value chains in their drive towards enterprise expansion. This leads to the concept of agricultural transformation.

Agricultural transformation can be described as the process of increasing the productivity of agriculture that results in the share of households dependent on agriculture for employment going down along with a reduction in the relative contribution of agriculture to the GDP (Timmer, 2014). Another useful definition is provided by Quinones and Diao (2011), "a process of sustainably modernizing agriculture and such a process is often measured by significant improvement in land and labor productivity, greater market-orientation and increased production diversification, as well as increased domestic and international competitiveness".

While agricultural enterprise expansion has to with increases in agricultural production and eventually increases in scales, agricultural transformation is not only about food production, it is also about the development of the economy. Obviously, agricultural transformation essentially encompasses the ingredients of sustainable and profitable enterprise expansion. Agricultural transformation process should to all intents and purpose bring about an agricultural sector that drives income growth and wealth for millions of farmers, facilitates achievement of food and nutritional security, and provides employment and competitively positions Nigeria into a leading player in global food markets.

Agricultural transformation strategy is not strange to Nigeria. In 2010 / 2011, the Government of Nigeria, began a reform of the agriculture sector through a strategy, called the Agricultural Transformation Agenda (ATA). The Federal Government through the Federal Ministry of Agriculture and Rural Development implemented transformation agenda through a set of inter-related programme interventions aimed at solving, in a holistic and integrated manner, the constraints and weaknesses that have for long constrained agricultural development. The ATA was designed to grow and develop agriculture as a business and consequently create jobs, ensure food security, promote private sector investments for wealth creation and ultimately maximize agriculture sector contribution to the country's economic growth. According to FMARD (2016), the Agricultural Transformation Agenda, ATA was in place from 2011 – 2015.

FMARD (2016) stated that ATA did not deliver on all the targets identified. "For example, Nigeria still imports about $3 to $5 billion worth of food annually, especially wheat, rice, fish and sundry items, including fresh fruits. As a result, Nigeria is not food secure. Wastage levels remain high in production areas, reducing supply of feedstock to processing factories, requiring them to keep importing supplies. The net effect is limited job growth across the agricultural value chain from input production to market systems, and continued use of limited foreign currency earnings to import vast quantities of food". As a result, the vision of the present administration for agriculture as contained in the Agriculture Promotion Policy (2016 – 2020) Building on the Successes of the ATA, Closing Key Gaps Policy and Strategy Document, is to work with key stakeholders to build an agribusiness economy capable of delivering sustained prosperity by meeting domestic food security goals, generating exports, and supporting sustainable income and job growth. As Nigeria embarks on the implementation of the Agriculture Promotion Policy (2016 – 2020) and barring the monster of policy reversals, let us focus on the ingredients of the shorter-term enterprise expansion.

Enhancing Enterprise Expansion

As the Nigerian population is currently put at 182 million and projected to hit 210 million by 2021 (NATIONAL POPULATION COMMISSION, n.d.), there is the compelling need to find a way to adequately meet rising demand for food while also meeting increased demand for other agricultural products such as biofuel feed stocks. To achieve this, the relevant stakeholders in the agricultural sector will need effect fundamental changes to the inherent and structural characteristics of agriculture in Nigeria. We must change the usual approach and perception of agricultural sector and ultimately make it a sustainably productive and profitable venture.

There are two broad strategies or approaches that human beings have had for affecting their food supply. The general approaches to increasing agricultural production are expansion in area under cultivation and boosting the output per unit of production (yield/hectare). While the former approach is referred to as extensive cultivation, the latter is referred to as intensive cultivation. Expansion of areas under cultivation is no longer an attractive option on account of population pressure and other unintended outcomes such as water pollution, biodiversity loss, and greenhouse gas emissions. According to FAO (2016), the expansion of agricultural land continues to be the main driver of deforestation. Consequently, there is a significant push toward sustainable intensification of agriculture that aims to optimize crop production per unit area while accounting for social, political, and environmental impacts.

Agricultural intensification can be technically defined as an increase in agricultural production per unit of inputs (which may be labour, land, time, fertilizer, seed, feed or cash) (Agricultural intensification, n.d.). Intensification occurs when there is an increase in the total volume of agricultural production that results from a higher productivity of inputs, or agricultural production is maintained while certain inputs are decreased (such as by more effective delivery of smaller amounts of fertilizer, better targeting of plant or animal protection, and mixed or relay cropping on smaller fields). Intensification that takes the form of increased production is most critical when there is a need to expand the food supply, for example during periods of rapid population growth. Intensification that makes more efficient use of inputs may be more critical when environmental problems or social issues are involved. A sustainable agricultural intensification plan should be able to balance economic, social and environmental issues.

According to FAO (2016), increased use of land, irrigation and agro-chemicals played a major role in

the growth of agricultural production during the Green Revolution. However, the gains of the Green Revolution were often accompanied by negative effects on agriculture's natural resource base, including land degradation, salinization of irrigated areas, the build-up of pest resistance and the erosion of biodiversity. Increasing agricultural production carries with it economic, environmental and social costs that may ultimately limit growth (Bender & Smith, 1997, Bennett et al., 2014). Essentially therefore, strategies to increase agricultural production and, by extension, enterprise expansion, should focus on increased production efficiency at lower environmental and resource costs. Examples include using improved irrigation techniques, increasing yield per unit input, climate-smart agriculture that produces less greenhouse gas per unit product, or other types of sustainable intensification.

While agricultural intensification is a preferred option for sustainable increases in agricultural production, it should be noted that sustainable agricultural intensification can only be supported by the adoption of new practices and technologies by farmers. According to FAO (2016), the key to sustainable agricultural growth is more efficient use of land, labour and other inputs through technological progress, social innovation and new business models. In the views of OECD (2011), and Troell *et al.* (2014), for agriculture and aquaculture to respond to future challenges, innovation will not only need to improve the efficiency with which inputs are turned into outputs, but also conserve scarce natural resources and reduce waste.

Opportunities in Agriculture

Further to the discussions on enterprise expansion, it necessary to make reference to the relevant areas in which the expansions can take place. This section presents the opportunities available in agriculture and in doing so goes a long way in underlining the point that agriculture goes beyond the cultivation of crops and rearing of animals. The following opportunities are available for people who are interested and qualified for employment in agriculture.

1. **Production Agriculture (Farming):** This involves the production of plants and animals for human beings and as raw materials for industries. Opportunities in production agriculture include operating farms as owners or renters. It also involves working as managers or as hired workers as is evident in commercial farms.
2. **Food Marketing:** Food marketing refers to all the activities and services that occur between the producers/providers of food and foodstuffs and the consumer. Food marketing obviously deals with what we eat. When we think of food, we should not just narrowly think of the farmer or the cattle herder. We should broadly think about the fisherman, the importers who bring in apple, oats etc. We should also think about what we eat that comes from industrial pants. These are all parts of the food production system. Examples of opportunities in the food marketing system include marketing of raw food (rice, beans, fish meat vegetables etc) and those engaged in the restaurant industry.
3. **Farm Service Marketing:** This marketing opportunity in agriculture involves all the activities and services that are brought to the farmer, herder, or producer in the form of purchased inputs. While in no way as large as food marketing system in terms of naira volume, the farm service marketing is equally important in the ultimate goal of getting food to the consumer.

Firms engaged in farm service marketing, agricultural production, and food marketing are collectively known as agribusiness.

SOLUTIONS AND RECOMMENDATIONS

For agriculture to be practised as a business, those engaged in it should be able to profit by it and increase both output and in scales. The following recommendations are therefore made.

1. Land is a vital factor for agricultural production and so guaranteed access by present and prospective farmers is critical. To this extent, government in whose hands is vest with the authority over ownership of land in Nigeria, should clearly define the conditions and rules guiding e the title and tenure over land for present and prospective farmers. Without secure rights to land farmers will not invest in improvements to their farms and to that extent limit enterprise expansion.
2. Access to inputs remains a challenge for achieving optimal productivity of agricultural practices. So, access to such inputs as fertilizer, improved seed, marketing and storage facilities should be guaranteed by the government on the one hand and public and private partnership arrangement on the other.
3. To increase agricultural productivity and enhance agribusiness, agronomic and other relevant information is regularly required by relevant stakeholders in the sector in the agriculture value chain for planning and decision-making.
4. Increasing productivity will also require significant investments in agricultural research and extension, in the road infrastructure that links farmers to markets and in the development of better rural infrastructure services such as health and education facilities.
5. Conscious efforts should be made by relevant stake holders especially the government to raise a new generation of young educated men and especially women farmers that are interested in the use of improved technologies and modern management approaches that help ensure farm profitability.

FUTURE RESEARCH DIRECTIONS

The agricultural sector is as old as Nigeria; however, it is still characterized by low productivity. There is the need to investigate why this so especially from the standpoint of policies and programmes enunciated to improve the agricultural sector in Nigeria.

CONCLUSION

Agriculture has an important potential influence on the economy o Nigeria. For the potentials of agriculture in an economy to be realized, the relevant stakeholders should realize that business as usual is not an acceptable option. Yield increasing and enterprise expansion inducing strategies should be implemented in both the short and the long term.

REFERENCES

Adesugba, M., & Mavrotas, G. (2016). *Delving Deeper into Agricultural Transformation and Youth Employment Nexus: The Nigerian Case*. IFPRI Working Paper 31.

Agricultural Intensification. (n.d.). Retrieved November 16, 2016, from http://www.fao.org/docrep/007/j0902e/j0902e03.htm

Agriculture for Impact. (2014) Small and Growing: Entrepreneurship in African Agriculture, A Montpellier Panel Report.

Agwu, N. M., Nwankwo, E. E., & Anyanwu, C. I. (2014). Determinants of Agricultural Labour Participation Among Youths in Abia State, Nigeria. *International Journal of Food and Agricultural Economics*, *2*(1), 157–164.

Amanor, K. S. (2010). Family Values, Land Sales and Agricultural Commodification in Southeastern Ghana. *Africa*, *80*(01), 104–125. doi:10.3366/E0001972009001284

Appropriation Act 2016. (n.d.). Retrieved January 1, 2017, from http://www.budgetoffice.gov.ng/pdfs/2016m/2016%20Appropriation%20Act.pdf

Bender, R., & Smith, M. (1997). Population, Food and Nutrition. *Population Bulletin*, *51*(4).

Bennett, E., Carpenter, S.R., Gordon, L.J., Ramankutty, N., Balvanera, P., Campbell, B., ... Spierenburg, M. (2014). Toward a More Resilient Agriculture. *Solutions: For a Sustainable and Desirable Future*, *5*(5), 65-75. Retrieved from https://www.thesolutionsjournal.com/article/toward-a-more-resilient-agriculture/

Bezu, S., & Holden, S. (2014). Are Rural Youth in Ethiopia Abandoning Agriculture? *World Development*, *64*, 259–272. doi:10.1016/j.worlddev.2014.06.013

Brooks, Zorya, Gautam, & Goyal. (2013). *Agriculture as a sector of opportunity for young people in Africa*. World Bank Policy Research Working Paper No. 6473.

Conway, G. (2012). *One Billion Hungry, Can We Feed The World?* Cornell University Press.

FAO. (2011). *Women in agriculture: Closing the gender gap for development*. Rome: FAO.

FAO. (2016). *The future of food and agriculture – Trends and challenges*. Rome: FAO.

Federal Ministry of Agriculture and Rural Development (FMARD). (2010). *Global Agricultural and Food Security Support Programme*. Nigeria: FMARD.

Federal Ministry of Agriculture and Rural Development (FMARD). (2016). *The Agriculture promotion policy (2016 – 2020) building on the successes of the ATA, Closing key gaps policy and strategy document*. Abuja, Nigeria: Federal Ministry of Agriculture and Rural Development.

Fox, M. L., Haines, C., Munoz, M. J. H., & Thomas, M. A. H. (2013). *Africa's Got Work to Do: Employment Prospects in the New Century*. International Monetary Fund Working Paper WP/13/201. IMF.

McMillan, M. S., & Hartgen, K. (2014). *What's Driving the 'African Growth Miracle'?* NBER Working Paper 20077.

National Bureau of Statistics (NBS). (2017). *Nigerian Gross Domestic Product Report, Q4*, 2016.

National Population Commission. (n.d.). *Data for national planning and development*. Retrieved November 16, 2016, from http://www.population.gov.ng/

New Partnership for Africa's Development (NEPAD). (2003). *Comprehensive Africa Agriculture Development Programme*. NEPAD.

Nin-Pratt, A. (2010). Technological Change and the Transformation of Global Agriculture: From biotechnology and gene revolution to nano revolution? In N. Ekekwe (Ed.), *Nanotechnology and Microelectronics: Global diffusion, economics and policy*. Information Science Reference.

Organisation for Economic Co-operation and Development (OECD). (2011). *Food and Agriculture. OECD Green Growth Studies*. Paris: OECD Publishing.

Quinones, E. J., & Diao, X. (2011). *Assessing Crop Production and Input Use Patterns in Ghana: What can We Learn from the Ghana Living Standards Survey?* GSSP Working Paper 24. Washington, DC: International Food Policy Research Institute.

Sumberg, J., Anyidoho, N. A., Chasukwa, M., Chinsinga, B., Leavy, J., Tadele, G., . . . Yaro, J. (2014). *Young People, Agriculture, and Employment in Rural Africa*. WIDER Working Paper 2014/080.

Timmer, C. P. (2014). *Food Security and Food Scarcity – Why it is Hard to End Hunger?* Pennsylvania University Press.

Troell, M., Naylor, R. L., Metian, M., Beveridge, M., Tyedmers, P. H., Folke, C., & de Zeeuw, A. et al. (2014). Does aquaculture add resilience to the global food system? *Proceedings of the National Academy of Sciences of the United States of America, 111*(37), 13257–13263. doi:10.1073/pnas.1404067111 PMID:25136111

Chapter 6
The Anchor Borrowers' Programme of Boosting Agricultural Production

ABSTRACT

On November 17, 2015, the government of Nigeria launched the Anchor Borrowers' Programme. The programme is aimed at boosting agricultural production and non-oil exports in the face of dwindling crude oil prices. Because the Anchor Borrowers' Programme is relatively recent and relevant to the main theme of the book—financing agricultural production expansion—its vision and mission are highlighted in this chapter with a view to informing and influencing the expected beneficiaries. The methodology employed is a systematic and analytic review of relevant literature. It is concluded that the Anchor Borrowers' Programme is a well-articulated initiative for economic linkage between smallholder farmers and reputable large-scale agro-processors with a view to increasing agricultural output and significantly improving capacity utilization of processors. It is recommended that the government resist the temptation of policies and programmes that are aimed at boosting agricultural financing and production rising and falling with the government that initiated them.

DOI: 10.4018/978-1-5225-3059-6.ch006

INTRODUCTION

This chapter focuses on one of the policies and programme of the Federal Government of Nigeria aimed at boosting agricultural production and non-oil exports in the face of dwindling crude oil prices. There many such programmes in Nigeria. This is highlighted here on account of its relative currency and relevance to the main theme of the book- agricultural production expansion. The chapter starts with the background information on the aims and objectives of the programme and the guidelines for operation. A review of the achievements of the programme so far achieved is included in this chapter recommendations and conclusion.

BACKGROUND

On November 17, 2015, the government of Nigeria launched the Anchor Borrowers'

Programme (ABP). The Programme is borne out of the consultations with such relevant stakeholders as the Federal Ministry of Agriculture & Rural Development (FMARD), State Governors, agro-millers, and smallholder farmers. The consultations were based on the need to boost agricultural production and non-oil exports in the face of dwindling crude oil prices and the negative effect on the revenue profile of Nigeria. According to Emefiele (2016), Nigeria's food import bills were exceptionally high; rice and wheat formed part of the items that consumed a whopping N1.0 trillion in foreign exchange. Emefiele (2016), further stated that food importation fueled domestic inflation, depleted our foreign reserves, displaced local production and created unemployment and concluded that import dependency especially on commodities of comparative advantage was neither acceptable nor sustainable.

In line with its developmental function, the Central Bank of Nigeria (CBN) established the ABP and is designed to create a linkage between anchor companies involved in the processing of agricultural commodities and the small holder farmers of the relevant key agricultural commodities. The ABP is aimed at the provision of farm inputs in kind and in cash to small holder farmers for the purpose of boosting production of the relevant commodities, stabilize supply of agricultural produce as inputs to agro processors and, as a result, address the country's negative balance of payments on food. The arrangement is such that at harvest, the small holder farmers supply their produce to the Agro-processor- the Anchor -who pays the cash equivalent to the farmer's account.

According to Central Bank of Nigeria (2016), the broad objective of the Anchor Borrowers' Programme is to create economic linkage between smallholder farmers and reputable large-scale agro-processors with a view to increasing agricultural output and significantly improving capacity utilization of processors.

Specifically, other objectives include:

1. Increase banks' financing to the agricultural sector;
2. Reduce agricultural commodity importation and conserve external reserves;
3. Increase capacity utilization of agricultural firms;
4. Create new generation of farmers/entrepreneurs and employment;
5. Deepen the cashless policy and financial inclusion;
6. Reduce the level of poverty among smallholder farmers; and
7. Assist rural smallholder farmers to grow from subsistence to commercial production levels.

LOAN BENEFICIARIES AND THE RELEVANT ENTERPRISES

For ease of administration, the loan beneficiaries are required to be in groups/cooperatives of not less than five and not more than twenty. Loan is targeted at smallholder farmers engaged in the production of the identified commodities. The loans are obtainable through eligible Participating Financial Institutions (PTIs). The PFIs are Deposit Money Banks (DMBs), Development Finance Institutions (DFIs), and Microfinance Banks (MFBs).

The relevant agricultural enterprises producing the named agricultural commodities on the basis of comparative advantage in the States are: Cereals (Rice, Maize, wheat etc.), Cotton, Roots and Tubers (Cassava, Potatoes, Yam, Ginger etc.), Sugarcane, Tree crops (Oil palm, Cocoa, Rubber etc.), Legumes (Soybean, Sesame seed, Cowpea etc.), Tomato, Livestock (Fish, Poultry, Ruminants etc.). Additionally, the Central Bank of Nigeria reserves the right to, from time to time, introduce to the programme any other commodity.

LOAN AMOUNT, INTEREST RATE LOAN TERM AND REPAYMENT

The per capita loan amount is arrived at from the costs and returns of production and agreed on by the relevant stakeholders. The pool of fund is provided from the N220 billion Micro, Small and Medium Enterprises Development Fund (MSMEDF).

The Interest rate for the loans under the Anchor Borrowers Programme is guided by the rate on the N220 billion MSMEDF. The interest rate, inclusive of pre and post disbursement, is currently at 9% per annum. The guidelines stipulate that the PFIs shall access the loan from the Central Bank of Nigeria at 2% interest rate and lend at a maximum of 9% per annum. The term of loans under the ABP shall be the gestation period of the identified agricultural enterprises that produce the agricultural commodities.

Under the ABP, loans granted to the small holder farmers shall be repaid with the harvested produce that shall be mandatorily delivered to the Anchor at designated collection point in line with the provisions of the signed contractual agreement. The produce delivered must cover the loan principal and interest.

THE ANCHOR

According to CBN (2016), the ABP guidelines stipulate that there shall be private large-scale integrated processors who have entered into a contractual agreement with the smallholder farmers to purchase the harvested produce at the agreed prices or as may be reviewed by the programme management committee. State Governments may act as Anchor upon meeting the prescribed conditions. The ABP guidelines document provides for input suppliers who shall submit expression of interest letter to the office of the Project Management Team for consideration and issuance of local purchase orders.

ADMINISTRATIVE MACHINERY

According to the CBN (2016), there shall be two models of administration of the ABP based namely on a. Private Sector-led and b. State windows. Each model has a Project Management

Team (PMT) which coordinates the implementation of the programme.

The Private Sector-Led Window

The Private Sector-led Window has the following set-up.

- The Chairman, (Head, Development Finance Officer, Central Bank of Nigeria (CBN)),
- Co-Chairman, (Representatives of Anchor Firms),
- Programme Manager, State Agricultural Development Programme (ADP),
- Representatives of participating Banks,
- Representatives of farmers associations, and
- Nigerian Agricultural Insurance Corporation (NAIC)

State Window

The State Window has the following set-up:

- The Chairman, (Head, Development Finance Officer, Central Bank of Nigeria (CBN)
- Co-chaired, a State Government Appointee,
- A representative of State Governments/Ministry of
- Agriculture and Rural Development/Agricultural
- Development Programme (ADP),
- Participating Banks,
- Anchor Firms,
- Nigerian Agricultural Insurance Corporation (NAIC), and
- Representatives of farmers associations.

MODE OF OPERATION OF THE ANCHOR BORROWERS' PROGRAMME

The process begins with the Anchor/State Government Writing Expression of Interest Letter to the Central Bank of Nigeria specifying the targeted agricultural commodities, proposed number of farmers, the hectares to be covered and the Participating Financial Institutions. The next is the formation of the Project Management Team. With the formation of the Project Management Team, the stage is set for the verification of the farmers and farm sizes. The verification is carried out by the Project Management Team.

The machinery of operation of ABN is further shaped by the confirmation of participation by the Head Offices of the Participating Financial Institutions. The identification of reputable agricultural inputs suppliers is undertaken by the Project Management Team. The Project Management Team further organizes Town Hall Meeting, where all the relevant stakeholder, including the input suppliers are in attendance to agree on such relevant issues as cost and returns of production per hectare, prices of the agricultural commodities, signing of agreement, etc. and any other relevant issues.

Next is the signing of tripartite agreement by the Participating Financial Institution, Anchor and the farmers. The signing of the tripartite agreement effectively gives rise to the submission of loan applications from head offices of the Participating Financial Institutions with the list of farmers in the prescribed format showing their accounts numbers, gender, farm size, Biometric Verification Number, telephone numbers, cooperative name and Local Government Areas. The farmers are also required to be registered on the National Collateral Registry (NCR).

TRAINING OF THE FARMERS

A compulsory training programme is provided for farmers participating in the ABP. The training programme covers such contents as farming as a business, improved agricultural practices, group management dynamics etc. Cost of such training shall be borne by the participating Anchor. However, partnerships with Development Partners are encouraged on the training of the farmers. Certificates issued at the end of the training shall constitute a requirement for farmers to access credit facility in kind and cash under the programme. Additionally, the Anchor/State Governments are required to provide agricultural extension services to complement the training ensure adherence to good agricultural practices and mitigate side selling.

COLLATERAL REQUIREMENTS UNDER THE ANCHOR BORROWERS' PROGRAMME

Participating farmers under the programme are required to deposit a minimum of 5% of the loan amount as equity in their accounts with the farmers' chosen Participating Financial Institution before loan disbursement is affected.

No input is accessed by farmer any who does not comply with the equity contribution requirement. The following are the collateral requirements acceptable under the programme: Cross guarantee by farmers in cooperatives/ groups, tripartite agreement signed by the parties, cross guarantee by farmers as registered on the National Collateral Registry (NCR), and equity contribution by the farmers.

RISK SHARING

According to CBN (2016), in order to engender participation of the Participating Financial Institutions in the programme, the Central Bank of Nigeria shall absorb 50% of the amount in default after satisfactory evidence that every means of loan recovery has been exhausted by the Participating Financial Institution in question. On its own part, the Participating Financial Institution shall bear the credit risk of the balance.

FUNCTIONS OF STAKEHOLDERS

Central Bank of Nigeria

The Central Bank of Nigeria (CBN) generally coordinates the entire Programme, serves as secretariat and Chair/Co-chair the Project Management Team (PMT). The CBN also provides the funds for the Anchor Borrowers Programme through the Micro, Small and Medium Enterprises Development (MSMEDF); and reviews the provisions of the guidelines as may be required.

Nigerian Agricultural Insurance Corporation (NAIC)

The NAIC provides insurance cover to the projects under the ABP; ensures timely processing and settlement of claims; and serves as member of the Project Management Team.

Development Partners

The Development Partners provides technical assistance to farmers, extension workers and banks; and may serve as member of the Project Management Team in partner States.

Participating Financing Institutions

The participating financial institutions verify eligible farmers and their farmlands; open account for the farmers; ensure due diligence on loan administration, monitoring and recovery; conduct searches on the National Collateral Registry (NCR) to ensure that none of the group member is in default of any other loan facility in any financial institution, mandatorily ensures that first applications for release of funds is accompanied by copies of executed tripartite agreements; register their interest in the collateral on the National Collateral Registry (NCR) as second and any subsequent applications for release of funds must be accompanied by evidence of complete registrations; sensitize the group members on the implication of the cross-guarantee as default by one member of the cooperative automatically puts all members in default irrespective of other members paying their individual loans; apply to the CBN for release of funds after completion of all the required conditions precedent to drawdown; credit individual farmer's account with the released funds within 5 working days; render monthly returns under the Scheme to the CBN in the prescribed reporting format; serve as member of Project Management Team; and carry out any other responsibilities as may be prescribed by the CBN from time to time.

Small Holder Farmers

The small-holder farmers shall organize themselves into groups/cooperatives; cross guarantee one another; demonstrate evidence of farm ownership/lease/rent

and agree to work with extension workers; utilize the facility (kind and cash) for the purpose for which it was granted; commit to abide by the terms of agreement and not to side sell produce; repay the loan as and when due by surrendering the output to the Anchor or State; provide equity contribution in cash of minimum of 5 per cent of loan amount; nominate a representative of the group/ association/cooperative to serve on the Project Management Team; ensure participating members open bank account and obtain Bank Verification Number (BVN).

State Government/Federal Capital Territory (FCT)

The State Government/FCT, as the case may be, shall co-Chair the public sector Anchor Borrowers Programme Project Management Team; submit

Expression of Interest to participate under the ABP; identify the two targeted agricultural commodities; provide extension services to all participating farmers; provide logistics support for the success of the programme including training for the farmers and extension services; establish a special 'farmers' court' to try defaulting parties; train identified farmers for participation under ABP; where State is an Anchor, must uptake the farmers produce and pay the farmers through the Participating Financial Institutions within 5 days at an agreed price; and carry out any other responsibilities as may be prescribed by the CBN from time to time.

Anchor Company

The Anchor Company shall co-Chair the Project Management Team of the private sector Anchor Borrower Programme; identify and organize farmers into groups/co-operatives; participate in the identification of input suppliers; train identified farmers for participation under the programme; provide extension service experts to support and ensure achievement of the targeted yield; monitor harvest and facilitate full evacuation of produce; establish produce collection centers; buy-up produce from farmers at agreed price; pay into farmers' loan account the money for the produce delivered within 48 hours after collection of produce; and carry out any other responsibilities as may be prescribed by the CBN from time to time.

Project Management Team (PMT)

The PMT shall coordinate project implementation; coordinate discussions on cost of production per hectares; communicate the decisions taken on the project to stakeholders; identify genuine input suppliers; coordinate and monitor project to ensure repayment; issue Local Purchase Order (LPOs); ensure timely distribution of inputs; report issues that cannot be resolved in the team to the relevant authorities; and ensure the programme is delivered and implemented as scheduled and within scope.

Nigerian Incentive-Based Risk Sharing System For Agricultural Lending (NIRSAL)

The NIRSAL shall collaborate on Technical Assistance.

INFRACTIONS AND SANCTIONS

The guidelines provide for sanctions against some specific infractions in the implementation of the ABP. The following infractions by the named stakeholders attract the specified sanctions.

For Participating Financial Institutions

Diversion of funds to unauthorized activities, attracts the following penalties: Amount diverted shall be recovered by the CBN; Penal charge at the maximum lending rate of the PFI on the amount diverted; outright ban from participating under other CBN Interventions following another infraction.

In the case of charging of unauthorized fees/interest, the sanctions are: reversal of the charged fees/interest; issuance of warning letter to the PFI; and outright ban from participating under other CBN interventions after two infractions.

For the infraction of charging of interest rates higher than that prescribed, the following sanctions are prescribed: Reversal of excess interest charged; penal charge at the maximum lending rate of the PFI; and issuance of warning letter to the PFI.

In the event of failure to disburse funds to the borrowers within specified period, CBN (2016), provides for: Penal charge at the maximum lending rate of the PFI; and recovery of the undisbursed amount plus interest.

For Anchor

Failure to collect certified quality output from farmers after going into agreement as the Anchor to the farmers attracts suspension of the Anchor from participation in the programme; and the Anchor will not be allowed to access agricultural and other CBN interventions. For failure to pay for collected commodities within the specified period, the following sanctions are stipulated: Anchor to bear the cost of accrued interest on the farmers' account from the due date; and call in the performance bond to settle for the produce delivered.

For Farmers

Side-selling attracts to the defaulting farmer(s) total prohibition from all CBN interventions; and blacklisting of the defaulter(s) from any intervention by the CBN; prosecution of the defaulter(s); and payment of the loan by the guarantors and cooperatives. With respect to input diversion as an infraction, the sanctions are: Blacklisting of the defaulters on any intervention by the CBN; and repayment of the loan by the guarantors and cooperative members. In the event of refusal to submit commodities to the Anchor, the sanctions are: Blacklisting of the defaulting farmers on any intervention by the CBN; prosecution of the defaulting farmers; and repayment of the loan by the guarantors and cooperative members.

For the default of fund diversion, the sanctions are Blacklisting of the SHF on any intervention by the CBN; prosecution of the SHF; and repayment of the loan by the guarantors and cooperative members.

For the Project Monitoring Team

For the infraction of insider related contracts and inflation of contract figures, the sanctions are: Suspension/prosecution of the culpable member(s); and reporting of the culpable member(s) to the relevant institution(s).

REVIEW OF THE ABN

According to Women Farmers (2017), The Federal Capital Territory Administration has as at June 2017, registered 9,000 farmers in the Federal Capital Territory (FCT) to participate in the CBN's Anchor Borrowers programme. It also reported that the achievements of the programme so far in the FCT include: increase in the average yield of rice from 2.0 metric tons/hectare to 3.5 metric tons per hectare; cassava yield witnessed upward review from 16 metric tons/hectare to 20 metric tons/hectare and the strengthening of the capacity of 2,000 farmers associations by way of training and linkage to agricultural finance.

According to Ujah and Komolafe (2017), the Central Bank of Nigeria, had disbursed N33.34 billion to 146,557 farmers under the Anchor Borrowers Programme. Details of the loan disbursement and repayment indicate that as at March 31, 2017, N33.34 billion had been released through 12 participating finance institutions in respect of 146,557 farmers across 21 states cultivating over 180,018 hectares of land. Of the total amount of N33.34 billion released, about N15.137 billion disbursed to 73,941 Kebbi State farmers have fallen due for repayment with N7.119 billion representing 47 per cent, repaid and returned to CBN.

According to Alli (2017), Manufacturers Association of Nigeria, MAN, has proposed the expansion of the scope of the Anchor Borrowers Programme to further boost production of key commodities and stabilize inputs supply to agro processors. MAN proposed that the scheme be expanded to include medium scale as well as large scale integrated processors to act as anchor companies, so as to increase private sector involvement in the scheme, and so boost production of key commodities, stabilize inputs supply to agro processors and address the country's negative balance of payments on food.

Delta Government has registered over 30,000 farmers across the selected enterprises of Cassava, Oil Palm, Rice and Fisheries for the Anchor Borrowers Programme. The aim is to boost the production of Fisheries, Cassava, Oil Palm and Rice production under the ABP in the state. (). It is further reported that for oil palm, government had commenced the raising of 220, 000 improved oil palm seedlings for distribution to 250 farmers for cultivation on 500 hectares of Oil palm plantations; and that five mini-oil mills have been fabricated and ready for distribution to farmers. It is expected that the oil palm programme will create 100 direct jobs and over 300 indirect jobs in the state. The report also stated that the annual fish production in Delta State has increased from 15,273 metric tonnes in 2008 to 24, 413 metric tonnes in 2016. The report further stated that the state government in 2015 introduced and distributed the new improved pro-vitamin A-Beta carotene cassava cuttings, sprayers, agro-chemicals, insecticides and fertilisers to farmers in the state.

According to *Bassey and Dipo (2017),* more than 36,000 farmers in Niger state benefiting from the Anchor Borrowers' Programme with 12,000 farmers joining the scheme last farming season, while the others joined in the 2017 cropping season.

In Akwa Ibom State that more than 17, 000 persons, mostly women and youths in rural areas have registered for the ABP. The first stage of the programme, which was the development of the ecosystem, mobilizing small holdings farmers and registration into cooperative society and opening of account with the recommended bank had been successfully carried out. Furthermore, the training of participant small-holder farmers has been carried out. It was further reported that the ABP may encounter some challenges in the state due to lack of such facilities as heavy duty equipment for clearing the thickly forested farmlands.

In Gombe State, the Central Bank of Nigeria has approved N660 million as loans for 3,479 farmers under the Anchor Borrowers Programme. The disbursement of the funds, which ought to have started earlier, was delayed due to multiple registrations by some farmers.

According to Peter (2107), participants in the Anchor Borrowers Programme in Cross River State have accused officials of the Bank of Agriculture, BoA, which is the participating financial institution in the State, of defrauding them under the programme. The participants reported that the officials were paying them N95,000 each, while being made to sign as collecting N250,000. Each farmer participant according to terms of the programme is to receive N250,000, and such farm inputs as rice seedlings, fertilize, etc. However, according to Peter (2107), the beneficiaries reported that after being told that the fertilizer was given at a subsidized rate, officials of the BoA gave them N95,000 each as their cash balance, after deduction for all the items, including fertilizer at N7,500 per bag, against the market value of N6,000. The state Commissioner for Agriculture explained that the per capita loan amount was N250,000 and that when the value of the accompanying farm inputs was deducted, the balance was given to each farmer.

SOLUTIONS AND RECOMMENDATIONS

The different stakeholders of the ABP should mandatorily set up monitoring and evaluation units in their different levels to collect collate and analyze achievements or lack of them based on the objectives of the programme. The resulting report should be placed in the public domain for more formal investigation into the achievements of the programme. Emphasis should be less on how much is approved, disbursed and eventually repaid than on how much the activities effectively addressed the mission and vision for which it was set up to accomplish. The government should endeavour to resist the temptation of policies and programmes that are aimed at boosting agricultural financing and production rising and falling with the government that initiated them. We should start thinking of ensuring continuity in our relevant and well thought-out agricultural programmes.

FUTURE RESEARCH DIRECTIONS

Formal empirical invstgation is recommended on the achievements of the Anchor Borrowers Programme in the direction of accomplishing the set objectives in general and increasing food production in particular.

CONCLUSION

The Anchor Borrowers Programme is a well articulated initiative to create economic linkage between smallholder farmers and reputable large-scale agro-processors with a view to increasing agricultural output and significantly improving capacity utilization of processors. Government should avoid the temptation of such a beautiful programme lasting only for as long as the government that initiated it.

REFERENCES

Alli, F. (2017). *MAN seeks expansion of CBN's Anchor Borrowers Programme*. Retrieved July 22, 2017, from http://www.vanguardngr.com/2017/07/man-seeks-expansion-cbns-anchor-borrowers-programme/

Bassey, O., & Dipo, L. (2017). *Nigeria: 36,000 Niger Farmers Benefit From CBN's Anchor Borrowers' Programme*. Retrieved July 22, 2017, from http://allafrica.com/stories/201707200245.html

Central Bank of Nigeria (CBN). (2016). *Anchor Borrowers' Programme Guidelines*. Abuja: Development Finance Department, Central Bank of Nigeria.

Emefiele, G. O. (2016). *Why we created the anchor borrowers' programme*. Retrieved July 13, 2017, from https://www.proshareng.com/news/.../Why... Anchor-Borrowers'-Programme-–.../33413

Peter, I. (2017). *Farmers allege fraud in CBN Anchor Borrowers Programme*. Academic Press.

Ujah, E., & Komolafe, B. (2017). *CBN disburses N33bn to farmers via Anchor Borrowers Programme*. Retrieved July 22, 2017, from http://www.vanguardngr.com/2017/05/cbn-disburses-n33bn-farmers-via-anchor-borrowers-programme/

Women Farmers. (2017). *9,000 FCT Register for Anchor Borrowers' Programme*. Retrieved July 22, 2017, from https://www.thisdaylive.com/index.php/2017/06/15/9000-fct-register-for-anchor-borrowers-programme/

Chapter 7
Financial Analyses

ABSTRACT

Financial analyses provide an account of an enterprise's past performance, a picture of its current financial strength/weakness, and a glimpse into the future financial potential. The owner of a farm firm periodically assesses and analyses the performance of the enterprise against predetermined objectives. The analysis is undertaken with the use of some tools. Because the discerning farmer needs to keep objectives of the enterprise in constant view, this chapter focuses on financial analyses with a view to highlighting the tools used in assessing a farm firm performance against predetermined objectives. The discussions are based on a review of relevant literature. The objectives of this chapter include defining the concept of financial analyses, describing the tools for analyses and showing worked examples of financial analysis, and highlighting the concept of rate of return. Financial analyses can be undertaken by management of an enterprise or by parties outside the enterprise. The three commonly used tools of financial analyses, namely the balance sheet, the income statement, and the cash flow statement, tend to be associated with some flaws, especially with regard to the reliability and validity of the accompanying figures. As a result, results from financial analyses should be interpreted with caution. It is further recommended that financial analyses be matched with a more observable measure of the financial health of a farm enterprise in terms of physical growth and expansion, credit worthiness, and how the enterprise meets its maturing obligations.

DOI: 10.4018/978-1-5225-3059-6.ch007

INTRODUCTION

The owner of a farm firm, to all intents and purposes, periodically assesses and analyses the performance of the enterprise against predetermined objectives. The analysis is undertaken with the use of some tools. The objectives of this chapter include define the concept of financial analyses, describe the tools for analyses and show worked examples of financial analysis.

BACKGROUND

Financial Analysis

Financial analysis, also known as financial statement analysis, involves maintaining, keeping and using records and relevant information to evaluate the financial performance of an enterprise. It is an attempt to look into the reported financial figures of a business outfit in order to determine its financial strengths and weaknesses (Anon, 2014). It is the process of identifying the financial strengths and weaknesses of the firm by properly establishing relationships between the items of a typical financial statement such as the balance sheet and the profit and loss account. A major objective of the financial analysis of farms is to judge how much farm families participating in a project will have to live on. Financial analysis can be undertaken by management of a firm or by parties outside a firm. The nature of financial analysis differs depending on the purpose of the analyst. For instance, trade creditors are interested in their claims being met by the firm over a very short period of time; their analysis will therefore be limited to the evaluation of the liquidity position. The suppliers of long-term credit, on the other hand, are interested in the firms' profitability overtime, its ability to generate cash to enable it pay interest, return their claims and the relationship between the various sources of funds. Similarly, shareholders are most concerned about the firms' earnings and as such concentrate their analysis on the firms' present and future profitability. Because management has the overall responsibility to see that resources of the firm are used most efficiently and effectively, it is concerned with every aspect of the financial analysis.

Analyzing the Performance of a Farm Enterprise

The owner of a farm firm periodically assesses the performance of the enterprise against predetermined objectives. The analysis is undertaken with the use of some tools. The most widely used financial statements are the balance sheet, the income statement and the cash flow statement.

In analyzing the performance of a typical farm enterprise, some indicators are used. The performance of the enterprise enables the analysts and relevant owners of the business to know if the business is expanding, contracting or outright stagnant with respect to growth. In effect, the indicators show the weakness or strength of the performance of the performance the business.

According to Olukosi and Erhabor (1988), indicators of the performance of an enterprise include measures of financial success, capital position, and farm size and resource use efficiency. Financial success, also referred to as the profitability of the enterprise, is determined by analyzing the net farm income statements. The capital position shows the solvency or otherwise of a firm. Solvency refers to the ability of the firm to meet its obligations are fall due. Measures of capital position are based on the analysis of the net worth statement. Measures of farm size are based on physical output of the farm, number of animals kept and/or crops cultivated, area of land under cultivation, total amount invested and total number of people employed. Efficiency indicates a level of performance that describes a process that uses the lowest possible amount of inputs to produce the greatest possible amount of outputs. It is defined as the quantity of output, Y, per unit of input, X, used in a production process. Resource use efficiency is obtained by looking at the ratio of output to input in the enterprise. These measures of financial performance of enterprises are focused on under the relevant financial statements.

MAIN FOCUS OF THE CHAPTER

Tools of Financial Analysis

The most widely used financial statements are the balance sheet, the income statement and the cash flow statement.

Balance Sheet

The balance sheet is a financial statement that summarizes a company's assets, liabilities and shareholders' equity at a specific point in time (Staff, n.d.) The three parts of balance sheet give investors and owners of the venture an idea as to what is owned and owed, as well as the amount invested by shareholders. The balance sheet also known as statement of financial position is a summary of the assets and liabilities of the business together with a statement of the owner's equity or net-worth. It shows the margin by which debt obligations would be covered if the business was terminated and all assets sold. It is a kind of still photograph of the financial state of the enterprise at a given moment; to this extent, it represents a stock concept as it refers to a particular point in time. A sample balances sheet for a rice farm enterprise is presented below.

As shown in Table 1, the items are listed in descending order of liquidity; the most liquid asset in the account is cash. Cash at hand refers to the rice output per hectare actually offered for sale multiplied by the unit market price. In standard balance reporting format, when an enterprise sells its goods or services on credit it is reported on the balance sheet as accounts receivable. Where some proportion of accounts receivable is not to be collected and, therefore, constitute a bad debt, or loss relevant, adjustments are made on the accounts receivable. The inventory for the rice farm enterprise is made up of the quantity consumed at home, the quantity for gifts and the quantity set aside for next year's planting. Summation of cash at hand, net accounts receivable, and inventories make up total current assets. In general, fixed assets include such items as farm tools and equipment that are of a fairly permanent nature and are employed in the normal running of the enterprise. The fixed assets are reported in the balance sheet at book value defined as the purchase price paid for the asset in question (i.e. the original historical cost) minus the cost of allowable depreciation at the point in time.

Land, a fixed asset, represented the largest single item on the balance sheet. The value in 2011 was ₦50,000. With the exception of land all fixed assets, are assumed to lose their economic value with time. So the value of land appreciated to ₦55,000 in 2012 and to ₦59,000 in 2013. In the case of land, their values which appreciated were adjusted to make them compatible with realistic estimates.

The values of total assets were ₦104,350 in 2011, ₦115,975 in 2012 and ₦119,400 in 2013. Total assets are a summation of current assets and fixed assets.

Table 1. Balance sheet of a rice farm enterprise as at 31 December of stated years

Assets	2011 (₦)	2012 (₦)	2013 (₦)
Current Assets			
Cash at hand	40,800	47,700	48,450
Net accounts receivable	200	200	200
Inventories	3,900	4,200	4,050
Total Current Assets	44,900	52,100	52,700
Fixed Assets			
Net Fixed Assets (Farm implements value less depreciation)	9450	8575	7700
Land (*)	50,000	55,000	59,000
Total Assets (T)	104350	115,675	119,400
Liabilities			
Current Liabilities			
Short term bank loans	5,000	6,000	6,500
Accounts payable	175	250	250
Accrued Expenses	875	875	875
Current Portion of Long term debt	250	300	325
Total current liabilities	6,300	7,425	7,950
Total Liabilities (L)	6,300	7,425	7,950
Networth (e)	98,050	108,550	111,450
Total Liabilities and Networth	104,350	115,675	119,400

(*) Real estate value adjusted to reflect market price in the locality
(e) = (T) minus (L)
Source: Adapted from Ejiogu, 2002

Under the liability accounts side of the balance sheet, the short-term loans come first. Liability accounts are used for the seasonal financing of current assets and activities of the enterprise. On the asset side, accounts receivable represents sales to customers made on credit by the enterprise; on the liability side, trade payable pertains to purchases on credit by the enterprise from suppliers. Accruals refer to specific direct and operating costs that a business outfit has expensed on which in fact have yet to be paid at the close of the accounting period. Examples are taxes, wages, rents etc. As the name implies, current portion of long term debt refers to current maturities of long-term liabilities.

Total current liabilities were ₦6,300 in 2011, ₦7,425 in 2012 and ₦7,950 in 2013. Networth which is the difference between the total assets and the total liabilities was recorded as follows: ₦98,050 in 2011, ₦108,550 in 2012 and ₦111,450 in 2013. The relatively high values of the owner equity show that the average rice farm is financed more from equity than debt. Furthermore, the prices appreciations were mainly derived not from increased production but from capital gains. However, capital gains provide a base for added borrowing for financing and expansion. It should be noted that the headings in the table are the major items and delineations that appear on most standard balance sheets. However, they are not the only entries that may appear on a balance sheet; some balance sheets may employ somewhat different terms and offer more details on their statements than others. Some accounts that are important for one enterprise may be minor or nonexistent for another enterprise.

In employing the net worth statement to determine the financial standing of a business, it is necessary to scrutinize the individual items in the statement. It should however be noted that the absolute size of the net worth does not tell the entire story. A clearer picture is obtained when the ratio of the total assets and total liabilities is examined in addition to the determination of how close to maturity the liabilities of the enterprise are.

As a rule of the thumb, current assets should, to all intents and purposes, should be able to match maturing current liabilities. Furthermore, to maintain sound financial health, short term loans should be channeled to short term investments and loan term loans to long term investments.

Ratio Analysis

In financial analysis, a ratio is used as an index for evaluating the financial position and performance of a firm. A ratio indicates a quantitative relationship which can be used to make a quantitative judgment. A ratio (quotient) is obtained when a number (dividend) is divided by another (divisor). Financial ratios can be classified into four categories namely: liquidity ratios, leverage ratios, activity ratios, and profitability ratios.

1. **Liquidity Ratios:** Liquidity ratios measure a firm's ability to meet current obligations. They relate to the firm's capacity to generate sufficient cash to meet its financial commitments as they become due. Liquidity ratios establish a relationship between cash and other current assets to current obligations. The cash flow statement is best used in analyzing liquidity of the firm. The ratios which indicate the extent of liquidity or lack of it are current ratio and quick ratio.

2. **Current Ratio:** Current ratio is obtained by dividing current assets by current liabilities. Current assets include cash and those assets that be changed into cash within a year and are included in current liabilities. These include creditors, bills payable and so on. For any firm, a ratio greater than one means that the firm has more current assets than current claims. Current ratio is a crude measure of liquidity because it does not take into account the individual components of the current assets.
3. **Quick Ratio:** Quick ratio is a more refined measure of a firm liquidity. It establishes the relationship between quick or liquid assets and current liabilities. The quick ratio is obtained by dividing the total of quick assets by total current liabilities. Quick assets = current assets – inventories. Inventories usually require sometime before they are realized into cash and are therefore considered to be satisfactory current financial condition. A quick ratio equal to, greater than or less than unity does not necessarily imply good or bad liquidity condition. This argument is plausible when it is considered that all book debts may not be liquid, and cash may be immediately needed to pay operating expenses.
4. **Leverage Ratio:** Leverage or capital structure ratios are calculated to determine the long-term financial position of a firm. These ratios indicate the funds provided by owners of the enterprise under consideration. Owners' equity in finance has a number of implications. First, debt is riskier than equity from the point of view of the firm. Second, employment of debt of advantage in two ways: they can retain control of the firm with limited stake; and that earnings will be magnified when the firm earns a rate higher than the interest rate on the invested funds.

However, leverage works in opposite direction. If the cost of debt is higher than the firm's overall rate of return, the earnings of shareholders will be reduced. Thus, use of debt magnifies the shareholders earnings as well as increases their risk. Thirdly, a highly debt-burdened firm will find it difficult to raise funds from creditors and owners in future. Leverage ratios may be calculated from the balance sheet items to determine the proportion of debt financing. They can also be calculated from the income statement items by determining the extent to which the firm has relied on debt funds in financing assets.

Leverage ratio calculation for the average rice farm enterprise for 2011-2013 is presented in Table 2. The leverage ratio is obtained by dividing the total liabilities by the net-worth for the different years. The values for the total liabilities and the owner equity are as shown in the balance sheet statement.

Table 2. Leverage ratio for a rice farm enterprise for 2011-2013

Year	Leverage Ratio
2011	0.05
2012	0.06
2013	0.06

Source: Adapted from Ejiogu, 2002

Table 2 shows that the leverage ratio for 2011 was 0.05 and in both 2012 and 213 it was 0.06 times the net-worth in the business. Stated in another way, the ratio of debt to equity ratio within the period was 6.94. Over the study period, the proportions of debt to equity were not only low but also showed that the small-scale farmer relied more on his equity for the farm operations. On account of this low leverage ratio which is a good solvency rating, the farmer can take on more debt to enable the farmer to expand and cultivate more hectares, use improved cultivation practices and procure and plant higher yielding seed varieties and even improve management skills.

Debt-equity ratio is the measure of relative claims of creditors and owners against the firms. It is the ease with which the proprietor can meet financial debts internally when and if the creditors demand. The debt-equity ratio is calculated by dividing the total debt (total liabilities) by the net worth. A very high ratio is unfavourable from the point of view of the firm. During the periods of low profits, a high debt-financial company suffers; it finds it difficult to pay the interest charges of creditors. A low debt equity ratio implies a greater claim of owner than creditors. The creditors view it as a satisfactory capital structure for a higher proportion of equity provides a larger margin of safety for them. When the profit is low, debt serving will prove less difficult for a firm with low debt-equity ratio. There is a need for balance between the use of debt and equity, the most appropriate combination being a tradeoff between return and risk.

Activity ratios evaluate the efficiency with which a firm manages and utilizes its assets. Also called turnover ratios, they indicate the rate at which assets are being turned over into sales. Activity ratio is the measure of the intensity of use of assets on the firm. One of the activity ratios is the inventory turnover.

$$\text{Inventory turnover } = \frac{Cost\ of\ goods\ sold}{Average\ \ inventory}$$

The cost of goods sold is obtained by subtracting closing inventory (stock) from the total opening inventory (stock) and the manufacturing costs (including costs of purchase). The average inventory is the average of the opening and closing inventories.

Profitability ratios measure the operating efficiency of a firm. Two major types of profitability ratios are calculated namely: profitability in relation to sales; and profitability in relation to investment.

If sales do not give enough profits, it would be difficult for an enterprise to cover operating expense and interest charges. Furthermore, if the firm does not generate sufficient return on investment, survival is jeopardized. One of the profitability ratios is the gross profit margin and is obtained by dividing the gross profit by sales.

Farm Income Statement

According to Lan (2012), the income statement reports how much revenue a company generated during a period of time, the expenses it incurred and the resulting profits or losses. Farm income statement also referred to as operating statement provides information on the gross receipts and total cost of production of an enterprise for a period of time usually a year. The income statement summaries the revenues and expenses of the enterprise during each accounting year and gives a kind of cinematic picture of activities overtime. It is a flow concept and therefore applies to a period of time as opposed to a point in time. This means that each value on an income statement represents the cumulative amount of that item for the accounting period under review.

Basically, the farm income statement is composed of three main parts namely: the total receipts (also known as gross receipts, total revenue or gross revenue), the total cost and the net income. Anon (2014), referred to the three parts of farm income statement as revenue, expenses and net income (loss).

A related term, net farm income, is the difference between gross receipts and total cost of production. Total cost of production comprises fixed costs and variable costs. Fixed costs are by definition the costs incurred within a production process on fixed factors which are not used up in the production process. Examples of fixed factors are plant, equipment, and land. By their nature, they are by degrees converted into outputs during production process. Variable costs are by definition the costs incurred within a production process on variable factors which are used up in the production process. Examples of variable costs include labour, seeds, fertilizer, etc. In the production of *okra* the cost of seed and fertilizer varies with the quantity of *okra* produced.

The gross receipts, also referred to as total revenue, total returns, or total value product is defined as the total output multiplied by the market price per unit of the product. Gross receipts are made of value of crops, livestock, poultry, fishery and their products, change in the inventory of crops, animals and their products, produce consumed at home and farm debts paid within the period (accounts receivable). An example of income statement is presented in Table 3.

Table 3 shows a rice farm income statement for three different years. The different sections of the net farm income statement should be noted. They are the total cash receipts, the total operating expenses also called the total

Table 3. Income statement for one hectare for the average rice farm enterprise 2011-2013

Items	2011	2012	2013
Receipts			
Rice sales (a)	40,800	47,700	48,450
Rice consumed at home	2,100	2,700	3,300
Increase in inventory	600	600	750
Total Receipts	43,500	51,000	52,500
Operating Expenses			
Hired labour	8,610	8,610	11,915
Rice seed	600	600	750
Fertilizer (a)	4,000	5,000	4,000
Agrochemical (a)	1,200	1,600	1,200
Miscellaneous (Feed of hired labourers, transportation cost etc)	7,200	8,200	9,000
Total operating expenses	21,610	24,010	26,865
Net Operating Income	21890	26990	25635
Fixed Expenses			
Taxes	175	250	250
Depreciation of farm tools	875	875	875
Interest on loans (*)	250	300	325
Total fixed cost	1,300	1,435	1,450
Total cost of production	22,910	25,435	28,315
Net farm income	20,590	25,565	24,185

(a) Valued at market price for the 3 different years.
(*) Calculated at 5% per month for 4 months
Source: Adapted from Ejiogu, 2002

variable costs, the total fixed cost, the total cost of production and the net farm income statement. These terms have been defined and the net farm income is obtained by subtracting the total cost of production from the gross farm income.

To ensure farm enterprise expansion, there should be sustained increases in the net farm income. On its part, increases in the net farm income can come about through increases in the total receipts or decreases in the total cost of production. Increase and improvement in the net farm income can come about by both increases in total receipts and decrease in total cost of production.

The total receipts, which is the gross output multiplied by the unit market price, is directly related to the farm output. So, enterprise expansion should primarily aim at increasing the farm output. However, for the enterprise expansion drive on account of the increased farm output to have meaning, the farm owner has to contend with the competitive nature of the market price of the product.

It is up to the farmer entrepreneur to explore effective ways of obtaining better prices for the products through cooperatives activities and engaging in contract farming. Better prices can be obtained from cooperatives that engage in the marketing of agricultural products. Contract farming can come in handy as a resource-providing contract in which a farmer obtains financing from a marketing outfit as well as guaranteed market outlet and production-management assistance. By definition, contract farming is a partnership-based business-oriented arrangement of forward agreements specifying the obligations of farmers (producers) and buyers.

It is a popular thinking that decreases in the total cost of production tend to increase the gross farm receipts. This is mathematically correct going by the equation:

NET FARM INCOME = TOTAL REVENUE- TOTAL COST

So any design that reduces the value of the total cost mathematically translates to a proportionate increase in the net farm income.

A decomposition of the total cost shows that it is composed of both fixed cost and variable cost. Because fixed cost cannot be changed within a production cycle, the proposed reduction in total cost with a view to increasing the gross farm income may not be targeted at the fixed cost in the short run. Therefore, the proposed cost reduction must be targeted at the variable cost items.

It is in the variable cost reducing considerations that the dexterity of the farm manager is mostly needed. For instance, if the agronomic instruction stipulates that 8 bags of fertilizer should be applied to a hectare of rice for optimum yield, it is ill advised for the farmer to cut on the recommended number of bags by way of cost reduction as it would eventually affect the yield. It is necessary to examine cost of variable inputs provided that the required quantities of inputs are available for optimum yield.

Robust and rigorous analyses of the net farm income with a view to measuring the financial success of an enterprise not only entail revenue increasing and cost reducing considerations. They also involve consideration of the returns to factors of production and the analysis of the gross, the operating and the fixed ratios obtained from the net farm income statement

Analyses of the Net Farm Income Statement

Analysts use the income statement to monitor revenues, expenses and profits and their trends over time. The absolute size and sign of the net farm income do not present a complete picture with respect to the financial success, or otherwise, of the enterprise. Further analyses are required beyond the examination of the value of the net farm income. The analyses include investigation of the returns to land, labour, capital and management. Also, a detailed consideration of the gross ratio, the operating ratio and the fixed ratio is required.

Analysis of Financial Relationships Based on the Farm Income Statement

Analysis of the financial relationships based on the entries in the farm income statement provides information on the returns or the income generating ability of the enterprise in question. Income statement ratios are usually used in analyzing the financial relationships as contained in the farm income statement. For more meaningful measure of the performance of the enterprise based on the farm income statement, the income statement ratios of a given enterprise can be compared to those of a similar farm in the locality. The income statement ratios can be alternatively compared to ratios of the same farm obtained from the farm income statement for some consecutive period of years.

Income statement ratios are divided into two namely expense to gross income ratios and income to investment ratios. Expense to gross income ratio establishes a connection between expenses to gross income. Income to investment ratio associates income to capital investments.

Let us illustrate how the analysis of the financial relationship based on the farm income statement works using the farm income statement in Table3.

Expense to Income Ratios

Ratios relating expense to income measure the margin with which the value of total returns exceeds the value of total costs. Controlling expenses in relation to income is a key to profitable farm operation (Lee *et al.,* 1980), and by extension enterprise growth and expansion.

One of the expense to income ratios is the operating ratio. It relates the variable expenses to the gross income. It is calculated using the formula:

$$\text{Operating ratio} = \frac{total\ variable\ expenses}{Gross\ income}$$

In the sample net farm income statement in Table 3, the total variable or operating expenses for 2011, 2012, and 2013 were N21,610, N24010, and N26,865 respectively. Similarly, the gross income (total receipts) for 2011, 2012, and 2013 were N43, 500, N51,000 and N52,500 respectively. Thus the total operating ratio amounted to 0.50 in 2011, 0.47 in 2012 and 0.51 in 2013. The average operating ratio was 0.49. An operating ratio of, for instance, 0.50 in 2011 means that the total operating expenses amounted to 50 kobo per naira of gross income.

The fixed ratio relates the fixed expenses to the gross income. It is calculated using the formula:

$$\text{Fixed ratio} = \frac{fixed\ expenses}{Gross\ income}$$

In the sample net farm income statement in Table 3, the total fixed expenses for 2011, 2012, and 2013 were N1,300, N1,435, and N1,450 respectively. The gross income 2011, 2012, and 2013 were N43,500, N51,000 and N52,500 respectively. Thus, the fixed expenses ratio for the three different years amounted to 0.03 each. Note that it is just a coincidence that the fixed expense ratio is the same for the different years. The average fixed ratio was 0.03. A fixed ratio of, for instance, 0.03 in 2011 means that the total fixed expenses also called overhead costs, amounted to 3 kobo per naira of gross income.

The operating and fixed ratios together make up the gross ratio. It is calculated using the formula:

$$\text{Gross ratio} = \frac{total\ expenses}{Gross\ income}$$

Gross Margin

Gross margin is defined as the difference between gross farm income and the total variable cost. Gross farm output is derived by summing up the output of the different enterprises in the organic unit of the farm firm. Total variable costs are the relevant costs consideration in the calculation of gross margin. Fixed costs are not considered in the calculation of gross margin.

Variable costs are relatively easy to assign to different enterprises of the farm firm, but not so for fixed cost. For instance, assume that a farmer pays interest on a loan for mixed cropping on a hectare of rented land. If the mixed cropping involves maize and cassava enterprises, how does the farmer charge the interest on loan and the rent on the land to the different enterprises? Allocation of fixed cost to respective enterprises tends to be arbitrary and therefore unrealistic. As a result, only the variable costs are essentially allocated to individual enterprises and subtracted from the enterprise gross outputs to arrive at the gross margin. As a tool of financial analysis, gross margin is used in determining the efficiency of farm enterprise with a view to making comparisons among enterprises. It comes in handy when fixed capital is a negligible proportion of the farm enterprise.

Gross Margin Analysis

Having defined the underlying concept of gross margin, we now focus on the related concept of gross margin analysis. The gross margin analysis entails the evaluation of the efficiency of named farm plans or enterprises for the purpose of making comparisons between the related farm plans or enterprises. As is stated in the discussions concept of gross margin, gross margin analysis is a relevant planning tool in circumstances where fixed capital is an insignificant size of the farm enterprise as is the case with small-scale agriculture. To better understand the use of gross margin analysis in the evaluation of farm enterprises, it is pertinent to attempt a definition of an enterprise.

According to Upton (1973), the different sections of a farm, each devoted to the production of one kind of crop or livestock are known as enterprises. In mixed farming and mixed cropping farming systems, farmers are faced with a wide range of alternative crops they could cultivate and livestock they could keep. Thus we can have yam enterprise, rice enterprise, cassava enterprise, poultry enterprise and so on. Enterprise combination is also possible and is prevalent in subsistence agriculture.

Gross margin analysis helps in determining the more or most, as the case may be, profitable enterprise or enterprise combination. For instance, if a farm business is made up of four enterprises namely: maize (m), cassava (c), melon (me), and yam (y), the total gross margin for the farm business is: $TGM = GM_m + GM_c + GM_{me} + GM_y$

$$GM_m = GI_m - TVC_m$$

$$GM_c = GI_c - TVC_m$$

$$GM_{me} = GI_{me} - TVC_{me}$$

$$GM_y = GI_y - TVC_y$$

Where GMm, GI_m, TVC_m, GM_c, GI_c, $TVC_{m,}$ GM_{me}, GI_{me}, $TVC_{me,}$ $GM_{y,}$, GI_y, TVC_y are gross margin for maize, gross income for maize, total variable cost for maize, gross margin for cassava, gross income for cassava, total variable cost for cassava, gross margin for melon, gross income for melon, total variable cost for melon, gross margin or yam, gross income for yam, total variable cost for yam respectively. The usefulness of gross margin analysis can be summarized as follows.

The Usefulness of Gross Margin Analysis

1. Lends itself to easy application in evaluating the profitability of subsistence farming system where the fixed cost component is negligible.
2. Gives a more valid result in evaluating the performance of different enterprises in a given farm using same capital items.
3. It serves as a useful guide in comparing and in selecting farm enterprises.
4. It helps in the investigation and the analysis of the variable cost component of a farm enterprise.

Gross Margin Analysis: Worked Examples

Let us use the data in Table 3 and make some modifications by way of showing an example of gross margin analysis.

Table 4 shows the gross margin for the rice enterprise for the three different years. The highest gross margin of N32990 was recorded for the year 2012; the least with the value of N26890 was recorded in the year 2011.

Note that the calculations are without the fixed cost component. If the different years were different enterprises, then comparison would be among the different enterprises. Let us illustrate the use of gross margin for comparison among different enterprises.

In Table 5, the gross margins for the three different enterprises are shown. If the analysis is to choose among the enterprises, the enterprise with the highest gross margin is selected. It should be stated that the sales figures are obtained by multiplying the total output offered for sale by the unit price of the product. Proportion consumed at home refers to the fraction of the total product devoted for consumption at home. To aid meaningful comparison, the analysis should also be based on specified dimensions of the plots or size of the enterprise. In this case, it is expressed in hectares.

Table 4. Gross Margin for one hectare of rice enterprise for the stated years

Items	2011	2012	2013
Cash Receipt			
Rice sales	40,800	47,700	48,450
Rice consumed at home	2,100	2,700	3,300
Gross Income (Total cash receipts)	43,500	51,000	52,500
Operating Expenses			
Hired labour	8,610	8,610	11,913
Rice seed	600	600	750
Fertilizer	4,000	5,000	4,000
Agrochemical	1,200	1,600	1,200
Miscellaneous (Feed of hired labourers, transportation cost etc	2,200	2,200	2,500
Total Variable cost of production	16,610	18,010	20,365
Gross Margin (Gross Income – Total Variable Cost of Production)	26890	32990	28315

Source: Adapted from Ejiogu, 2002

Table 5. Gross Margin for one hectare each of yam, maize and cassava enterprises

Item	Yam	Maize	Cassava
Cash Receipt			
Sales	40,800	47,700	48,450
Proportion consumed at home	2,100	2,700	3,300
Gross Income (Total cash receipts)	43,500	51,000	52,500
Operating Expenses			
Hired labour	8,610	8,610	11,913
Planting materials	600	600	750
Fertilizer	4,000	5,000	4,000
Agrochemical	1,200	1,600	1,200
Miscellaneous (Feed of hired labourers, transportation cost etc)	2,200	2,200	2,500
Total Variable cost of production	16,610	18,010	20,365
Gross Margin (Gross Income – Total Cost of Production)	26890	32990	28315

Source: Adapted from Ejiogu, 2002

Cash Flow and Cash Flow Analysis

A cash flow statement is a listing of the flows of cash in and out of a business or project within a specified time. In business, cash inflows ordinarily arise from these activities - financing, operations and investing. Financing cash flows are cash received from the issue of debt and equity instruments, or paid out as dividends, share repurchases or debt repayments. Operational cash flow refers to cash received or expended as a result of the internal business activities of an enterprise. A typical example is cash earnings. Investment cash flows consist of money received from the sale of assets, or spent on capital expenditure. Examples are investments, acquisitions and long-life assets. Cash outflows result from expenses or investments.

Cash flow shows the amount of cash generated and used by an enterprise in a given period. Cash flow can be presented as a record of transactions that took place in the past, such as the sale of a particular product or service. Cash flow can also be projected into the future, representing what an enterprise expects to receive and to spend. As a listing of cash flows that occurred during a past accounting period, cash flow is regarded as a cash flow statement.

A projection of future flows of cash is called a *cash flow budget*. By creating a cash flow budget, the sources and uses of funds for future time periods can be determined. Cash deficit periods are identified beforehand and corrective actions promptly put in place to manage the deficit. Management of the perceived deficit may involve changing the timing and varying the volume of relevant transactions. Management may also involve such considerations as whether to borrow where to borrow.

Cash flows convey information on both size and timing of cash inflows and outflows. It is common to find cash flows for a year in which the month end and the year end balances are all specified.

The difference between cash inflows and cash outflows is called the net cash flow. A net cash flow can either be positive or negative. A positive net cash flow occurs when an enterprise receives more money than it is spending within a specific period of time. A positive net cash flow enables an enterprise to pay its bills as they fall due. A negative net cash flow implies that the enterprise is receiving less cash than it is spending within a specific period of time. As a result, the enterprise may need to borrow money to pay its current obligations.

Cash flow analysis is a systematic investigation into how a business operation is running, from where its money is being generated, and how the money generated is being spent. Working capital, defined as the amount of money needed to facilitate business operations and transactions, is an important part of a cash flow analysis. Working capital is calculated as current assets less current liabilities. Calculation of the amount of working capital provides a quick analysis of the liquidity of the enterprise over the relevant accounting period. For example in the balance sheet in Table 1, the working capital for the year 2011 is the total current assets (made up of cash at hand and crops for sale) N44,900 minus total current liabilities (made up of taxes, short term loan and depreciation charges) N6300 giving a value of N38,600.

In absolute terms, this figure for working capital is big and positive. However, it makes more meaning when the amount is related to the amount of money needed to facilitate business operations and transactions for a time period. If the amount is sufficient, then the business enterprise is liquid.

It should be underlined that the balance sheet, the income statement and the cash flow statement are all tools of financial analysis. The balance sheet or net worth statement is a stock concept which refers to a point in time and shows the solvency of a business at a particular point in time. The balance sheet shows the value of assets and liabilities and the net worth being the amount by which the value of assets exceed the value of liabilities. The net worth reflects the value of the investment in the business by the owners of the business. The farm income statement is a flow concept and refers to period of time. It contains records of income and expenses within an accounting period. A cash flow statement shows *liquidity* while an income statement shows *profitability of a business enterprise.*

Rate of Return

A return is the gain or loss of an enterprise or a security in a specified period of time. Rate refers to the measurement of the speed or the number of times something happens or exists during a particular period. Rate of return can be construed as the gain or loss on an investment over a defined time period. Rate of return is usually expressed as a ratio or a percentage. In general, there are three different types of return measures: return on investment, return on equity and return on assets. Each one is essentially calculated the same way, but the inputs have different labels (Staff, 2016).

Return on Investment

Return on investment measures the amount of gain or loss on an investment relative to the cost of the investment. The return on investment is positive if the return is a gain and negative if it is a loss. Return on investment is obtained by dividing the return of the investment by the cost of the investment. Return on investment in the commonest way to calculate the profitability of an investment.

The return on investment formula:

$$\text{Return on Investment} = \frac{Income\ from\ Investment - Cost\ of\ Investment}{Cost\ of\ Investment}$$

In the above formula, if the cost of investment is greater than the income from investment, the rate of return on investment is negative and vice versa.

The net farm income statement easily lends itself to the calculation of return on investment; where the net farm income divided by the total cost of production (or the total variable cost as the case may be) multiplied by a hundred gives the return on investment for the period in question (usually a year).

Return on Equity

Return on equity, RoE, is another generally employed measure of return used in interpreting business performance. It is used to determine the return on the capital supplied by the owners of the enterprise in question. For example, for the net farm income of the enterprise in Table is N20,590 for the year 2011, and the equity capital (net worth) of the enterprise over the same time as recorded in Table 1 is N98050, the return on equity is 21%. This means that the enterprise makes a return of 21% or 21 kobo on every naira of equity invested.

Return on Assets

Another popularly used measure of return is the return on assets, RoA. Measurement of return based on RoA has its focus on the assets of the named enterprise. The RoA for a named enterprise for a named period, usually a year, is obtained by dividing the net (farm) income by the value of total assets. For example, if net income for the rice enterprise for the year 2011 in Table 3 is N20,590, and total value of assets for the enterprise over the same time period as is shown in Table 1 is equal to N104,350, the return on assets is N20,590 divided by N104,350, or 19.73%.

SOLUTIONS AND RECOMMENDATIONS

A major objective of the financial analysis of farms is to judge how much farm families participating in a project will have to live on. A major drawback in the use of financial analysis is the possibility of working with padded and unreliable figures. As a result, farm financial statements should be read with caution. Furthermore, a more observable measure of the financial health of a farm enterprise should be the physical growth and expansion of the enterprise, the credit worthiness of the enterprise and how the enterprise meets its maturing obligations.

FUTURE RESEARCH DIRECTIONS

Further research should be directed towards understanding the relation between results of financial analysis and the enterprise expansion of the entity for the financial analysis refers.

CONCLUSION

The owner of a farm firm, to all intents and purposes, periodically assesses and analyses the performance of the enterprise against predetermined objectives. Financial analysis can be undertaken by management of a firm or by parties outside a firm. The analysis is undertaken with the use of some tools. The tools may be associated with some flaws especially as regards the reliability of the accompanying figures. As a result, results from financial analysis should be interpreted with caution.

REFERENCES

Anon, . (2014). *Accounting and Finance for Business Analysis*. DELTACPE LLC.

Ejiogu, A. O. (2002). *Financial analysis of small-holder rice production enterprise in Imo State* (Unpublished doctoral dissertation). University of Nigeria Nsukka.

Lan, Z. J. (2012). *Introduction to financial statement analysis.* Retrieved October 13, 2017, from http://introduction-to-financia-statement-analysis.pdf

Olukosi, J. O., & Erahbor, P. O. (1987). *Introduction to Farm Management Economics Principles and Applications*. AGITAB Publishers Zaria.

Staff, I. (2016, June 1). *Rate of Return*. Retrieved November 16, 2016, from http://www.investopedia.com/terms/r/rateofreturn.asp

Staff, I. (n.d.). *Balance Sheet*. Retrieved November 16, 2017, from http://www.investopedia.com/terms/b/balancesheet.asp

Chapter 8
Measures of Project Worth

ABSTRACT

Investments usually involve the procurement of assets for which using marginal analysis may not be adequate in evaluating their worth to an economic activity in an enterprise. Furthermore, all the costs involved in the purchase of fixed assets are not ordinarily charged to the account of a production period. It is against his background that this chapter focuses on the concept of measures of project worth with a view to enabling farmers to obviate related problems in capital budgeting, non-discounted measures of project worth (pay-back period, average rate of return), discounted measures of project worth, benefit-cost analysis, net present value, decision criteria in using the net present value, internal rate of return, calculating the IRR, and interpreting the IRR. Discussions were based on a review of related and relevant literature. Conclusions and recommendations are made based on the discussions.

INTRODUCTION

Measures of project worth basically involve aggregating and comparing the costs and the benefits of a project to determine the values of the project as a whole. These measures of project worth are mainly divided into two broad categories namely discounted measures of project worth and undiscounted measures of project worth. Discounted measures of project work take into consideration the time value of money while undiscounted measures do not. Measures of project worth are typically employed to determine which among various projects is to be accepted or to be rejected for investment purposes.

DOI: 10.4018/978-1-5225-3059-6.ch008

Investments usually involve the procurement of assets for which adopting marginal analysis may not be suitable for evaluating its worth to an economic activity (Drummond & Godwin, (2004). It is impossible to buy or use a marginal unit of a mechanical device e.g. a tractor or a grinder as such an asset comes in all-or-nothing packages. Investment in fixed assets to all intents and purposes involves production periods in the long run. Costs involved in the purchase of fixed assets are not ordinarily charged to the account of a production period. That is the justification for the asset to decline in value, over the expected useful life span, due to wear and obsolescence. Furthermore, where the analysis involves choice between projects that will last several years and also have variously shaped costs and benefits streams, then simple direct comparison of the costs and benefits is less than accurate.

Because evaluating costs and returns on the margin is both impractical and inappropriate for all-or-nothing packages and because of time dimensions in the value of money, simple difference between total costs and total benefits is anything but satisfactory in the measurement of project worth. We must therefore determine and compare the present value of the expected costs and the present value of the expected benefits to determine the accurate measure of a project worth.

The determination of the present value today of some future value is, by definition, called discounting. So the determination of the worth of any project that employs the computation of the present value of some future value is referred to as using discounted measures of project worth. However, there are some measures of project worth that do not employ the discounting procedure. These are commonly referred to as non-discounted measures of project worth.

In the course of running a farm, investments have to be made. For the investments to be profitable, the investment decision must be based on valid analysis of the costs and returns. According to Dueling and Sherill (2006), decisions involving the commitment of present funds in exchange for future fund are referred to as investment decision. This chapter highlights the tools that are essentially used in reaching such investment decisions by the investing farmer with a view to improving profitability and enhancing enterprise expansion. The rest of the chapter is divided into Non-Discounted Measures of Project Worth and Discounted Measures of Project Worth. It is concluded with recommendations.

Non-Discounted Measures of Project Worth

Non-discounted measures of project worth refer to those tools which do not take into account the time value of money that are employed in evaluating investment proposals. These non-discounted measures of project worth use the absolute values of the costs and benefits of projects in evaluating decisions on their choice or rejection. Examples of measures of non-discounted measures of project worth include pay period, and accounting rate of return.

Pay-Back Period

The pay-back period can be defined as the number of years required to recoup the initial cash outlay in a project. The pay-back period is derived by dividing the initial cash outlay by the annual cash inflow. Assuming a project has an initial cash investment of N100,000 and annual cash inflow of N20,000 for 8years, the payback period is calculated thus:

$$\text{Payback Period} = \frac{Initial\ Cash\ Investment}{Annual\ Cash\ Inflow}$$

$$= \frac{N100,000}{N20,000}$$

= 5 years

The payback period for this hypothetical project is 5 years. This project can be considered as viable and therefore acceptable. It would have been unviable and therefore unacceptable if the payback period is equal to or greater than 8 years. Where there is an expected or required payback period, the calculated payback period is compared to the acceptable payback period. If the calculated payback period is less than the maximum acceptable payback period, then the investment proposal is accepted. On the other hand, if the calculated payback period is more than the maximum acceptable payback period, then the investment proposal is rejected. In the event of unequal cash inflows, the calculation of the payback period is obtained thus: Assuming that the annual cash inflows were distributed as follows: Year1 N39,000; Year 2 N35,000; Year 3 N35,000; Year 4 N47,000; Year 5 N55,000; Year

6 N30,000; Year 7 N40,000; and Year 8 N39,000. To calculate the payback period, add the cash inflows to get a figure close to the original cash outlay of N100,000; thus in the first two years, N74,000 of the original investment will be recovered, followed by N35,000 in the third year. But it remains N26, 000 to recoup the N100,000 mark, So the payback period is obtained thus: 2 years + (N26,000/N35,000), or 2. 74years.

The major drawback of the payback method as a measure of project worth is that it does not take into account cash flows after the payback period and as a result cannot be reliably employed in measuring the profitability of a project. Furthermore, the playback period method considers only the recovery period as a whole and not the size and timing of the cash flows.

Average Rate of Return

The average rate of return is an accounting method of evaluating the expected profitability of a project. It takes into account the tax and the depreciation values associated with the project. It is obtained by finding the ratio average annual book earnings after tax to the average investment in the project. This is represented by the following equation:

Average rate of return =

$$\frac{\textit{Average annual book earnings after tax} \quad X\ 100}{\textit{Average investment in the project}}$$

Assume that a project has an initial investment cost of N100,000 and the following annual book earnings: Year1 N39,000; Year 2 N35,000; Year 3 N35,000; Year 4 N47,000; Year 5 N55,000; Year 6 N30,000; Year 7 N40,000; and Year 8 N39,000. Assume a 50% tax rate and depreciation, on straight line basis, of N10,000. The average rate of return is calculated thus:

Average earnings before tax and depreciation =N40000

Depreciation (given) =N10,000

Net earnings before tax =N30,000

Tax at 50% rate =N15,000

Net earnings after tax =N15,000

$$\text{Average rate of return} = \frac{Average\ annual\ book\ earnings\ after\ tax X100}{Average\ investment\ in\ the\ project 1}$$

$$= \frac{N15,000}{N100,000} \quad \frac{X100}{1}$$

=15%

In this example, the average rate of return for the project is 15%. This is then compared with the expected or required rate of return to reach an investment decision. Note that in the example, the initial coast outlay occurred once. Where the cost outlay occurred more than once, the average would then be used. In terms of drawbacks in the se of average rate of return, it can be observed that this method does not take into account the timing of the annual earnings.

Discounted Measures of Project Worth

Discounted measures of project worth refer to those measures of project worth which take into consideration the time value of money. The time value of money is graphically encapsulated in the aphorism that a bird in hand is worth two in the bush. Based on the concept of time value of money, the present value of one naira today will definitely be more than one naira any day in the future. If one has one naira as at present and wants to figure out how much it is worth a year in the future assuming an interest rate, the future value will definitely be more than one naira. This is obtained by compounding the cash flows to their future value. Conversely, one naira in the future is obviously less than one naira in the present. This is obtained by discounting the cash flows to their present value. To better appreciate the concepts of present value and future value of an investment, let us focus closely at the equally related concepts of compounding and discounting.

Compounding

By definition, compounding is the process of determining the future value of an investment given a present value, an interest rate per time period and that interest is earned on interest. By way of illustration, suppose that N100 is deposited in a bank account that pays 5% compounded annually. At the expiration of the first year, the account will yield N105 made up of N100 which is the initial deposit and N5 interest. At the expiration of the second year, the account will yield N110.25 made up of the following:

Initial deposit =N100

Interest on N100 in Year1 =N5

Sub-total at end of Year1 =N105

Interest on N105 in Year2 =N5.25

Grand total end of Year2 =N110.25

Mathematically, the yield at the end of year 1 is

FV_1 =N100(1+0.05)

=N105

For the yield at the end of year 2, it is obtained thus

FV_2 = N100$(1.05)^2$

=N110.25

To derive a general equation for compounding to obtain the future value of an amount in the present, the foregoing calculation can be generalized as

$$FV_n = PV_o (1+r)^n$$

Where FV_n = future value of a deposit at the end of n years

$PV_{o=}$ present value /initial deposit

r= interest rate

n = number of time periods in years

The general equation for compounding is directly applicable where the calculation is on annual basis. There is a slight modification where the interest rate is paid for periods other than annually for instance quarterly of semiannually. Thus the general equation for compounding for the future value at the end of year *n* for an investment given a present value where the interest rate is paid *m* times in a year we have

$$FV_n = PV_o\left(\frac{1+r}{m}\right)mn$$

Where FV_n = future value of a deposit at the end of n years

$PV_{o=}$ present value /initial deposit

r= interest rate

n = number of time periods in years

m= number of times interest is paid in a year

Discounting

By definition, discounting is the process of determining the present value of an investment given a future value, a discount rate per time period. It can also be understood as running the compounding process in the opposite direction. To see if we will arrive at the same figure, let us look at the compounding problem in a reverse order. Suppose you want to buy a book worth N105 next year and that a deposit bank pays interest of 5% on a year deposit. How much must you deposit in the bank in order to realize N105 one year from now? Solving for PV in the equation for FV we have

$$FV_n = PV_o\,(1+r)^{n}$$

$$N100 = PV\,(1+0.05)^{1}$$

$$PV_o = \frac{N105}{(1+0.05)^{1}}$$

= N100

This means N100 will be deposited now in order to have N105 one year from now at 5% interest rate. In other words, N100 is the present value of N105 due to be received after a year at the interest rate of 5%. Thus the equation for the general rule for discounting is:

$$PV_o = \frac{FV_n}{(1+k)^n}$$

Where PV = present value /initial deposit

FV_n = future value at nth year

k= discount rate

n = number of years

The general equation for discounting is directly applicable where the calculation is on annual basis. There is some slight modification where the interest rate is paid for periods other than annually for instance quarterly of semiannually. Thus the general equation for calculating the present value when the discount is calculated more than once in a year is derived as:

$$PV = \frac{FV}{\frac{1+k}{m}^{mn}}$$

Where PV = Present value

FV_n = Cash flow at the end of year n

k= discount rate

n = number of time periods in years

m= number of times interest is paid in a year

Benefit-Cost Analysis

Benefit cost analysis is essentially a form of economic analysis of an investment or policy alternative in which the estimated total benefits of the proposed investment or policy alternative are compared with the estimated total cost. As a discounted measure of project, all future costs and benefits are reduced to the present value using an appropriate discount rate. According to Horne (1980), profitability index or benefit-cost ratio of a project is the present value of future net cash flows divided by the initial cash outlay. The equation for benefit-cost calculation can be expressed as:

Where A_t is the cash flow for period t, k is the discount rate and n is the terminal period in which cash flow is expected. A_0 is the initial cash outlay or cost which occurs at time 0.

Where costs are incurred in periods other than the initial cost outlay, the equation is modified thus:

Where B_t is the returns or benefit for period t, i is the discount rate and n is the 1,2,3,4,... n number of years in which the returns or benefits are expected. C_t is the cost for period t, and n is the 1,2,3,4,... n number of years in which the costs expected.

The benefit and cost streams for the n number of years are discounted to their present value. The ratio of the present value of the benefits and costs is then determined. If the benefit cost ratio is equal to or greater than one, then the project is viable.

Net Present Value

Net present value (NPV) also called net present worth (NPW) is the discounted value of the benefits less the discounted value of costs of a project. The foregoing definition can be presented in the following equation:

$$Net\,Present\,Value = \sum_{t=1}^{t=n} \frac{B_t - C_t}{\left(1+\mathrm{r}\right)^{\mathrm{t}}}$$

Where Net Present Value = NPV, B_t = discounted value of benefits, C_t =discounted value of costs, r= interest or discount rate, t = 1,2,3,4..., n number of years. For independent projects, accept the project with NPV o zero or greater.

As a discounted measure of project worth, calculation of net present value is closely related to the discounting process which is in itself is the determination of the present value of an investment given a future value, a discount rate per time period.

Recall that the equation for discounting is:

$$PV_o = \frac{FV_n}{(1+k)^n}$$

Where PV $_=$ present value /initial deposit

FV $_n$ = future value at nth year

k= discount rate

n = number of years

This formula is employed when calculating net present value of an investment. For example, assuming a cash outlay of a project to be N18,000 and annual benefit streams for the next five years of N5,000; N4,500; N6,000; N5,500; and N6,650. Assuming a discount rate of 5%, the net present value is calculated thus:

NPV= PV of Benefits minus PV of Cost (Cash Outlay)

NPV=

$$\frac{5000}{(1+0.05)^1}+\frac{4500}{(1+0.05)^2}+\frac{6000}{(1+0.05)^3}+\frac{5500}{(1+0.05)^4}+\frac{6500}{(1+0.05)^5}-18000$$

= 4762 + 8367.30 + 16339.20 + 19502.45 + 28141.75 -18000

=77112.7-18000

=N59112.30

In addition to solving the equation mathematically, the present value of the stated benefit at the end of the specified year can be obtained by looking up the discount factor under the given discount rate in a table of discount factors. The relevant discount factor is then multiplied with the stated benefit to obtain the present value. In this example, the cash outlay o N18,000 occurred once and at the beginning of the project. That is the reason the amount was not discounted. On the other hand, if the cash outlay occurs more than once and also in the future, the present value o such costs should be calculated in the process of determining the net present value of a project.

The flow of benefits and costs may be even and equal in different periods or uneven and unequal in different periods. In the example where the annual benefit streams for the next five years are N5,000; N4,500; N6 000; N5,500; and N6,650, the benefit steams are clearly uneven and unequal in the different time periods. When the benefit streams are even and equal for the stated number of years, otherwise called annuity, the calculation of present value and ultimately, net present value, is easier. In calculating the present value of even and equal stream of benefit, the annuity factor under the stated discount rate and specified terminal year is looked up in the table. The annuity factor is then multiplied with the benefit expected to be received in each period. For example, assuming a cash outlay of a project to be N18,000 and the equal and even annual benefit stream for the next five years N5,000; and assuming a discount rate of 5%, the net present value is calculated thus. In the table of annuity, look up the discount factor for annuity at 5% for five years. This is 4.3295. The discount factor for annuity at 5% for five years, 4.3295, is multiplied with the annual equal and even stream of benefit, N5000. That is:

N5000X 4.3295=21647.5

The NPV = 21647.50-18000

= 3,647.5.

Decision Criteria in Using the Net Present Value

Using the net present value method, accept independent project whose NPV is positive or zero. Conversely, reject independent project having negative NPV. If there are two or more mutually exclusive projects, accept that with highest NPV. It can be construed as the rate of interest which brings the present value of future benefits to be equal to the present value of future cost so that the net present value of the investment is zero.

Internal Rate of Return

The Internal Rate of Return, (IRR) can be construed as the rate of growth a project is expected to generate over a period of time. IRR is also referred to as the marginal efficiency of capital, is the rate of return at which discounted present value of net cash flow of a project is equal to the initial cost of the investment. It is the rate of interest (discount rate) which makes the present value of future benefits to exactly equal the present value of future cost so that the present value of the whole investment is zero. Use of this discounted measure of project worth requires a determination of the rate of interest which equates the present value of the future cash flows with the initial capital cost of the project. Clearly the problem here is to calculate the rate of interest at which the discounted present value of future cash flows is equal to the cost of the project.

It should be pointed out that to solve for the rate of interest rate, the following data should be supplied:

1. The initial outlay of the project;
2. The total lifespan of the project;
3. The magnitude of future cash flows in each period; and
4. Table of discounted values.

Mathematically, IRR is represented by the equation:

$$\sum_{t=0}^{n}\left[\frac{A_t}{\left(1+r\right)^t}\right]=0$$

Where A_t is the cash inflow or cash out flow for period t, n is the terminal period in which a cash inflow or cash out flow occurs, the sigma sign represents the sum of discounted cash flows at the end of the period n.

An investment with an appreciably higher IRR value than other available alternative offers a better chance of strong growth. However, the actual rate of return that a given project eventually generates often differs from the estimated value of the IRR. The IRR method of project worth has a clearly defined decision rule: accept projects that whose IRRs are higher than the required rate of return. In investment planning, it is usual to establish a required rate of return (RRR) which defines the minimum acceptable return in percentage terms that the proposed investment is expected to earn in order to be considered profitable. If the IRR is less than the required rate of return the project is rejected because embarking on such a project at such a rate would definitely decrease the shareholders' wealth.

Calculating the IRR

To illustrate, assuming that an investment has initial cost of N3000 and yields N30,000 20years later. To calculate the IRR, we can proceed thus:

The NPV is given by NPV = $\frac{N30000}{(1+r)^{20}} - N3000$

If the NPV is zero, then $\frac{N3000}{(1+r)^{20}} = N3000$

Therefore $(1+r)^{20} = \frac{30000}{3000}$

$1+r = \sqrt[20]{10}$

r =.12201845- 1 1

= 0.1220145

Hence r= 0.122 or 12.2%

Thus the internal rate of return necessary to equate the present value of the cash flows with the present value of the outflows is 12.2%.

Except that by a shot in the dark that hits the target, it is difficult to choose the discount rate that equates the present value of the expected cash with the present value of the expected inflows. As a result, it is fashionable to resort to the procedure of trial and error to compute the discount rate that equates the present value of benefits equal to the value of initial cash outlay. Let us consider another example, this time an annuity problem.

$$8000 = \frac{2000}{(1+r)} + \frac{2000}{(1+r)^2} + \frac{2000}{(1+r)^3} + \frac{2000}{(1+r)^4} + \frac{2000}{(1+r)^5} + \frac{2000}{(1+r)^6} + \frac{2000}{(1+r)^7}$$

The foregoing equation can be solved using a suitable computer programme. Solving manually, it should be understood that the cash flow stream is represented by an annuity of N2000 for 7 years. We are required to solve for the discount rate in the stated number of years that when multiplied by N2, 000 equals the initial cash outlay of N8000.

In order to minimize the number of trial and error attempts in finding the accurate discount rate, we can proceed by dividing the initial capital cost, c, by the value of annual cash flow, a:

$$\frac{c}{a} = \frac{8000}{2000} = 4$$

If the initial capital cost of an investment is divided by the value of the annual benefit stream, it gives the present value of each naira of the initial capital cost that is required to satisfy the condition that the discounted value of the benefits is equal to the initial capital cost. In the foregoing example, the present value of N1 per annum for seven years is equal to 4.

The next step in the calculation is to determine from annuity table which rate gives the nearest figure to N4 for seven years. The annuity table contains the present value of one naira per annum receivable for n years at a rate of interest r. The nearest figures from the annuity table for n= 7 are:

16% 4.039

17% 3.918

The discount rate must therefore be between 16% and 17%. Using interpolation, which is the process of calculating a desired value between two other values, we have:

Present value at the discount rate of 16% = 4.039

Present value at the discount rate of 17% = 3.918

Difference between the two discount factors = 0.121

Present Value at 16% = 4.039

Required discount rate =4.000

Difference between the two = 0.039

The rate therefore lies $\frac{0.039}{0.121}$ x 1 away from 17%

=17- 0.3223140495867769= 16.67768595041322

= 16.68%.

According to Gittinger (1994), there is a rule guiding the procedure for interpolating the value of the internal rate of return. The rule is summarized in the following statement:

$$\text{Internal rate of return rates} = \text{Lower discount rate} + \text{Difference between the discount rates}$$

$$\frac{\begin{matrix}\text{Present worth of}\\ \text{incremental net benefit}\\ \text{stream }\left(\text{cash flow}\right)\text{ at}\\ \text{the lower discount rate}\end{matrix}}{\begin{matrix}\text{sum of the present worth}\\ \text{of the incremental net bene}-\\ \text{fit streams }\left(\text{cash flows}\right)\\ \text{at the two discount rates,}\\ \text{signs ignored}\end{matrix}}$$

$$\frac{\textit{Present worth of incremental net benefit stream (cash flow) at the lower discount rate}}{\textit{sum of the present worth of the incremental net benefit streams (cash flows) at the two discount rates, signs ignored}}$$

Using the foregoing statement involves the use of some arithmetic procedures. As a result, it can be referred to as the arithmetic method of calculating the value of IRR.

To choose the discount rate which makes the net present value of benefit equal to the net present value of cost, a systematic procedure of trial and error is involved. It is recommended that the spread between the higher and the lower discount rates should not be more than 5 percentage points.

Computation of internal rate of return notoriously involves trial and error and some bit of luck. Deterring the relevant discount rates is a bit problematic. It is advised that the HDR be the first to be determined. It is also suggested that the HDR be the rate at which the NPV is negative. Let us demonstrate the use of the foregoing arithmetic method formula in calculating the value of IRR.

Suppose that a project has an initial cost outlay of N10,000. The project is further associated with cost of N4,000 each and yields of N8,000 each year for the next three years. Calculate the IRR. We first determine the HDR. This can be determined through trial and error until the rate with negative NPV is obtained. Let us work with 10% as the higher discount rate with a negative NPV. Based on the recommendation that the spread between the higher and the lower discount rates should not be more than 5 percentage points, let the HDR be 8%.

Note that the NPVs at the different discount rates must be used in the calculation. We then calculate the NPV of the project at 8% discount rate.

The present values of costs at 8% discount rate are calculated as follows:

1. The N10,000 is the initial cost outlay and occurs in the present; so the present value is still N10,000.
2. The values of cost for the three years occur in equal annual series of N4,000. To obtain the present value of the series, we use the annuity factor 2.5771. (This figure is read off at the table for present worth annuity factor under 3 years at 8%). Multiplying the annual series of cost (N4000) with the annuity factor gives us the present value of cost: N4000 X 2.5571 = 10308.4. Add the initial cost outlay to obtain the total present value of cost: 10308.4+ 10000 = N20308.4.

The present value of revenue at 8% discount rate is obtained as follows: The values of revenue for the three years occur in equal annual series of N8,000. To obtain the present value of the series, we use the annuity factor 2.5771. (This figure is read off at the table for present worth annuity factor under 3 years at 8%). Multiplying the annual series of revenue (N8000) with the annuity factor gives us the present value of cost: N8000 X 2.5571 = 20616.8. This is the total present value of revenue.

The NPV of the project at 8% discount rate is the difference between the present value of revenue and the present value of cost: NPV = N20616.8-N20308.4=N 308.4.

The net present value of the project at 10% discount rate is calculated in a similar manner. Hence the present value of cost is N10,000 which is the initial cash outlay, plus N4000 X 2.4869= N9947.6+ N10000= N19947.6 (The figure 2.4869 is read off at the table for present worth annuity factor under 3 years at 10% discount rate). The total present value of cost is N19947.6. The present value of revenue at the discount rate of 10% is N8,000 X 2.4869= N19895.2.

The NPV of the project at 10% discount rate is the difference between the present value of revenue and the present value of cost: NPV = 19895.2-N19947.6=N -52.4.

Substituting the relevant values into this formula

$$Internal\ rate\ of\ return\ rates = Lower\ discount\ rate +$$
$$Difference\ between\ the\ discount\ rates$$

$$\frac{\begin{array}{l} Present\ worth\ of\ incre- \\ mental\ net\ benefit \\ stream\ (cash\ flow)\ at \\ the\ lower\ discount\ rate \end{array}}{\begin{array}{l} sum\ of\ the\ present\ worth \\ of\ the\ incremental\ net\ bene- \\ fit\ streams\ (cash\ flows) \\ at\ the\ two\ discount\ rates, \\ signs\ ignored \end{array}}$$

we have:

$$\text{IRR}= 8\% + 2\% \left(\frac{308.4}{360.8}\right)$$

$$= 9.71\%$$

Interpreting the IRR

The higher the IRR of an investment proposal, the more acceptable it is to invest in the project. As a result, the project with the highest IRR is usually considered most acceptable and undertaken first. This is usually applicable for mutually exclusive projects.

The IRR is usually interpreted based on an established criterion. This criterion can be referred to as the required rate of return. In investment analysis, it is customary to establish a cut off rate or a required rate of return to ascertain the acceptable minimum percentage return that the proposed investment must earn in order to be worth investing in. In other words, a long as the IRR is higher than the rate at which funds are obtained for investing in the project, the investment proposal is acceptable.

IRR is helpful in interpreting the acceptability of an investment proposal by comparing the profitability of establishing new operations with that of expanding old ones.

SOLUTIONS AND RECOMMENDATIONS

The accuracy of comparing costs and benefits is a function of the integrity of the values used and the skills of the analysts. It is recommended that farmers should be able to evaluate the profitability of their proposed investments, where the relevant skills are lacking it is advisable that professionals be invited to undertake the analysis.

FUTURE RESEARCH DIRECTIONS

The computation of IRR for use as a measure of project worth tends to be arduous. There is therefore the need for future research directions to focus on making manual calculations of IRR less tasking.

CONCLUSION

Measures of project worth basically involve aggregating and comparing the costs and the benefits of a project to determine the values of the project as a whole. These measures of project worth are mainly divided into two broad categories namely discounted measures of project worth and undiscounted measures of project worth. Discounted measures of project work take into consideration the time value of money while undiscounted measures do not.

REFERENCES

Drummond, H. E., & Godwin, J. W. (2004). *Agricultural Economics* (2nd ed.). Prentice Hal Inc.

Dueling, T. N., & Sherill, W. W. (2006). *Entrpreneurism: Exploring entrepeneurism from a business process perspective*. Atomic Dog Publishing Cincinnati.

Horne, J. V. (1980). *Financial Management and Policy*. Prentice/Hall International London.

Chapter 9
Farm Planning and Analysis

ABSTRACT

Nigeria's renewed interest at transforming agriculture essentially entails that the process of farm planning should be in constant focus at all levels of agricultural enterprise production. As agricultural enterprise becomes more business oriented and environmental regulations become more site-specific, farm planning becomes more useful. It is against this backdrop that this chapter focuses on the concept of farm planning and analysis with a view to equipping the investing farmer with the rudiments of farm planning and analysis. The methodology is based on a systematic review of related and relevant literature. It is recommended that farmers should be able to constantly evaluate the profitability of their proposed investments; where the relevant skills are sufficiently deficient, it is advisable that professionals be invited to undertake the analysis.

INTRODUCTION

The need for planning and analysis of farm enterprises to keep track of profitability and expansion of the enterprise cannot be overemphasized. For one thing, it enables the farmer and the relevant stakeholders to know the strengths and the weaknesses of the farm enterprise. For another, it provides information about the enterprise for the relevant stakeholders on the opportunities for expansion and investment.

Farm planning entails the process of allocation and organization of scare farm resources in such a way as to achieve the overall objective of a farm

DOI: 10.4018/978-1-5225-3059-6.ch009

enterprise. It involves organization and management of scarce resources to realize the specified goals on a continuous basis. It is essentially a decision-making and choice-making process in terms of resource use and enterprise combination in the face of scarcity. Farm planning is a decision-making process to the extent that it involves the organization and management of limited resources to realize the specified goals of the farm enterprise. Farm planning is choice-making to the extent that, in the face of all possible alternatives, the farmer chooses what to produce, how to produce, when to produce and where to buy and sell all aimed at achieving the objective of a farm enterprise in the most profitable way. Farm planning involves consciously selecting the most profitable course of action from among all possible alternatives.

The objectives of a farm enterprise which farm planning aspire to achieve include economic (provision of basic human food and fiber needs), environmental, (enhancement of environmental quality and the critical resource base on which agriculture actually depends), and social (enhancement the quality of life for farmers and society in general). A farm enterprise may be pursuing one objective or a combination of objectives. Some of these farm objectives create trade-offs, where decisions to enhance one tend to detract from another. According to Janke (2000), farm planning helps integrate the objectives into a comprehensive and achievable whole, creates specific action steps and a time-line for reaching each objective, and benchmarks for monitoring when unified objective has been achieved.

Farm planning may basically be for a short or for a long term. Short term planning lasts for a relatively short period usually for a year or less. However, there may be overlaps in time horizons of both the short and the long terms. This comes about when there is a combination of both aspects of short term and long-term planning. For instance, when it is required to establish the best combination of an annual crop to grow under a permanent crop, the agronomic instructions and technology packages for annual crops are duly modified. The modifications are in line with the year-to-year changes in the development of the permanent crop.

It is needful to define such related terms as feasibility study, business plan, and farm plan and thereby bring out their differences and similarities. The differences and similarities are expected to enable the investing farmer appreciate which is applicable in what situation.

According to Women in Business (2010), feasibility studies refer to valid and verifiable information about a business structure, the products and services, the market, logistics of actually delivering the product or service, the resources needed to make the business run effectively and all relevant information about the business. The term can also be construed as a controlled process for identifying problems, opportunities, defining objectives, describing situations, defining successful outcomes, and assessing the range of costs and benefits associated with several alternatives for solving the identified problems (Thompson, 2005). A feasibility study, also called business opportunity analysis (Marzec, n,d), is an analysis of how successfully a project can be completed accounting for factors that affect the project such as economic, technological, legal, social and environmental factors (Staff, 2015). Project managers use feasibility studies to determine the potential positive and negative outcomes of a proposed project before committing resources into the project. According to Mesly (2017), the goal of feasibility study is to place emphasis on potential problems that could occur should the project be pursued and eventually determine whether the project should go ahead or else be abandoned altogether. Generally, feasibility studies precede technical development and project implementation. Feasibility studies evaluate a project's potential for successful outcomes; therefore perceived objectivity is an essential factor in the credibility of any feasibility studies for potential investors and lending institutions.

To the financial institutions and venture capitalists, business plan has different definitions (NUC, n,d). Banks tend to consider formal business loan applications which they put out to be a reflection of a relevant business plan. Venture capitalists see them as investment proposals, purely fund-raising documents. Corporate managers think of them in terms of departmental budgets and financial forecasts (Touchie, 2005). According to Kuratko and Hodgetts (1998), the business plan describes to investors and suppliers of finances all of the events that are likely to affect the proposed venture. Details are required for various projected actions of the venture, with associated revenues and costs outlined. A Business Plan describes a business opportunity. It is like a road map because it shows what to expect and what alternative routes that can be taken to arrive at the expected destination. According to Blasingame (2011), a business plan is the result of thinking, research, strategy and reaching conclusions about how to pursue opportunities.

Business Plan enables one to determine what commitment that needs to be made to the venture (Department of Trade and Economic Development, 2010), and when such commitment should be made. According to Dueling and Sherill (2006), business plan is a road map for starting and running a business; it is also used to convince bankers, venture capitalists, family, friends and even oneself to invest in a specific business.

A farm plan is used if it is intended to develop a piece of land for agricultural purpose (Department of Economic Development, Jobs, Transport and Resources, 2017). It brings out the natural resource management features of the land, the use to which the land can be put and the impacts of the proposed development on the natural resources and the community. In the views of Blaisingame (2011), a plan is like a parked car; planning is taking that car on a trip. Planning is measuring a business motion against the baseline of assumptions and projections made in the plan. It allows one to see how the business idea stands up to external forces. Thus a business plan is important and planning is essential.

Marzec (n.d), suitably summarized the similarities and the differences between feasibility study and business plan and stated that a feasibility study is a planning tool similar to a business plan. The feasibility study is done to flesh out the possibilities in an initial business proposal; the business plan then fully describes the business proposal and its financial projections. Feasibility studies answer the question, will this work? A business plan answers the question: how will this work? Marzec (n.d), further stated that feasibility studies are for the entrepreneurs benefit to determine if the business is worthwhile. On the other hand, business plans are targeted at investors, lenders to explain how the business works. Feasibility studies usually compare several possible scenarios on how a business might work. Business plans describe a specific business and include different sections which are of interest to the audience. In a similar vein, NUC (n.d), stated that developing the business plan is one of the best ways to define the blueprint, strategy, resource, and individual requirements for an on-going venture. NUC (n.d) also stated that when undertaking a new venture (feasibility studies), or remaking an old one (business plan), a written strategy can help ensure its success and that business plans and feasibility studies are required as controlled process for identifying business problems and opportunities, determining objectives, describing situations, defining successful outcomes, and assessing the range of costs and benefits associated with several alternatives for solving a problem.

The chapter specifically highlights the need for farm planning and analysis and their relevance in enabling the profit-oriented farmer keep track of the profitability of farm enterprises and ultimately enhance enterprise expansion. The chapter specifically focuses on the basic ways of planning an investment using the instrument of budget. It defines the concept of farm budget, looks at the definition of partial budget, and complete farm budget. The chapter is divided into the following headings: Importance of Farm Planning, Whole Farm Planning, Tools for Farm Planning, Cash Flow Budgeting, Linear Programming, Conclusion and Recommendations.

IMPORTANCE OF FARM PLANNING

Nigeria's renewed interest at transforming agriculture essentially entails that the process of farm planning should be in constant focus at all levels of agricultural enterprise production. For instance, the government Agricultural Transformation Agenda of 2011 – 2015 emphasized the need for Nigerian agriculture to shift focus towards how a self-sustaining agribusiness focused economy could be built. Additionally, the new policy and strategic focus, The Agriculture Promotion Policy (2016 – 2020), is now on how to build on the initial progress made, and transition Nigeria to a new plane in terms of agribusiness performance (FMARD, 2016).

As Nigerian agriculture shifts focus towards self-sustaining going concern, funding requirements become larger, lenders and investors become more likely to require that farmers have business and marketing plans, cash flow projections and updated balance sheets. As agricultural enterprise becomes more business oriented, and environmental regulations become more site-specific, farm planning becomes more useful. It is against this backdrop that the following importance of farm planning is presented.

1. Farm planning is important in the transformation of the agriculture sector in Nigeria to a new plane in terms of agribusiness performance. As a road map, a well-articulated farm plan enables the farmer to get to where the farmer wants to go. The destination can be in terms of accomplishing, in a more organized manner, the objectives of the farm enterprise.
2. Farm planning helps in tracking the costs and returns of farm. The tracking of the costs and returns not only throws light on the profitability of the farm enterprise but also shows the different components of the cost items. In the event of gain, loss or even breakeven, the farmer or the interested analysis will be able to determine the source the outcome and take the necessary action by way of consolidation or remediation.
3. Similarly, farm planning aids the farmer in easily identifying the input and investment needs. The quantity and quality of input needed in the farm enterprise are essentially dictated by the decision-making process

of farm planning. Sources and uses of investment funds and enterprise expansion are borne out of the farm planning process. A good farm plan helps in determining the creditworthiness of the farmer in relation to particular enterprise.

4. One of the greatest importance of farm planning is that it enables efficient allocation of scarce resources to their best alternative uses.

WHOLE FARM PLANNING

Farm planning is usually associated with the concept of whole farm planning. It is therefore necessary that we connect the two terms. When a holistic approach is taken in farm planning to the extent that adequate consideration is given to all farm assets over a long-time period of up to 50years or more in line with all possible aims and objectives the farmer may aspire, whole farm planning is in focus. On account of the long-time horizon, adjustments and re-planning along the way are a part of whole-farm planning. Whole farm planning is essentially a process.

Whole farm budget is a process of planning, property design and management based on natural resources and economic factors (Department of Economic Development, Jobs, Transport and Resources, 2017). It encompasses the knowledge and skills required to plan a sub-division, irrigation layout, assess land capability and potentials of a farm; it includes such aims of the farmer as profit, life style, family wellbeing, sustainability of production and all. In the views of Janke (2000), whole farm planning means taking the total of a farm's enterprise budget, and joining them into one plan or budget. Thus, all fixed and variable costs are allocated to an enterprise budget, and the composite of the enterprise budgets comprise the financial outlook for the farm as a whole. Economic whole-farm planning also can include financial goal setting.

Farm planning, be it whole farm planning or short-term planning, is a dynamic process. It should be pointed out that farm planning results in a reference document, a road map for the future of the enterprise, a business plan or better still a farm plan. The inevitable refinement and modification of the farm plan in the course of time as dictated by the inherent biological characteristics of agricultural enterprises define the dynamic process of farm planning.

Figure 1. Farm planning process

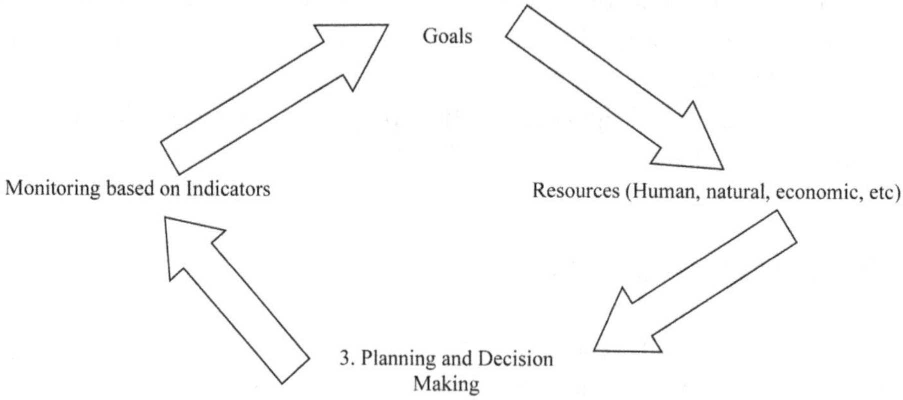

A typical farm planning process as shown in Figure 1 is guided by at least four interlinked phased cycle of goal setting, resource assessment, decision-making, indicator-based monitoring and refinement/re-enforcement of goals. Basically, the planning process is guided by the farm enterprise vision, mission and goals to clearly understand 'where it is', effectively project 'where it wants to be' and successfully strategize on 'how to get there'.

The goal setting phase involves developing perspectives and long-term vision for the desired quality of life for one and one's family within the community, the future of the farm enterprise, how the farm enterprise will provide the needed income and living environment (Minnesota Institute for Sustainable Agriculture (MISA) 2011). Valid and reliable assessment of the resources available to achieve the set goals is also needed at the second phase of the cycle. The resources in question include natural, human, financial resources. Others include capital assets, and crop and livestock management practices. In the third phase of the farm planning process, it is necessary to identify and assess management alternatives and thereby develop and implement an action plan. The fourth and final phase, after developing an action plan that is compatible with the goals set, is to monitor progress toward these goals.

Farm planning is only as good as the effort, time and information that go in to it. By way of keeping the plan on track and focused on the goal, the cycle of assessment, planning, and re-assessment can be periodically repeated as dictated by relevant indicators. The relevant indicators are usually made up of the expected and the actual costs and revenues flows over the planning horizon. Scarcely do things turn out exactly as planned; so periodic adjustment, and where necessary complete overhaul may be required to keep the plan in the desired direction.

TOOLS FOR FARM PLANNING

There are some tools employed in the farm planning process. The two basic planning tools available are budgeting and linear programming. Enterprise budgets show which parts of the farm operation are making a profit and which parts are not. Whole-farm budgets are best at analyzing the complementarity among different parts of a farm. In both cases, reliable and valid record keeping is essential in keeping track of the progress made by an enterprise in achieving the set objectives. Let us take a closer look at the different tools.

Farm Budgeting

Farm budgeting as a farm planning technique may be defined as the process of generating detailed physical and financial statement of a farm enterprise or of an expected modification in farm enterprise over a specified period of time. The physical and financial statement is generally expressed in monetary terms through the estimation of receipts, expenses and profit. Farm budgeting is the technique of analyzing plans for the employment of factors of production at the control of the farmer. Farming budgeting gives rise to a farm budget.

According to Dueing and Sheril (2006), a budget can be described as a plan set out in monetary terms, put together and approved before a defined period of time and usually depicting either or both the expected income to be generated and the expenditure to be incurred during the specified time period. According to them, one of the ways through which enterprises strive to achieve their strategic objectives is through preparing budgets and monitoring actual performance against the set objectives.

There are various approaches to budgeting including incremental budgeting and zero budgeting. In incremental budgeting, the precious year's figures provide the basis for the current year. Zero budgeting begins anew with each budget plan to the extent that each area of cost and revenue is new at the beginning of the budgeting period.

Farm planning with the associated use of the budgeting tool is successful to the extent of the availability, validity and reliability of the records that go into them. It cannot be overemphasized that for agricultural enterprises to be run as a business, there is the urgent need for the farmers to keep farm records especially in view of their special role in the process of farm planning.

The following are the different types of farm budgeting.

Partial Budgeting

Partial budgeting refers to estimation of the cost and returns of a section or a component of an enterprise. It is a planning tool employed for the estimation of monetary gain or loss that may arise in changing one or two aspects of a business or a farm enterprise. It is employed when contemplating making a partial change to the enterprise in question.

Partial budgeting gives rise to a partial budget which shows the expected change in profit for a proposed change in the farm enterprise. Partial budgeting is usually employed when there is a proposed modification in a farm plan. As a result, a partial budget specifically contains the costs and returns items that are expected to change in the event of the implementation of the proposed modification. The estimated the total increase in returns minus the estimated total reduction gives the net change in returns. Positive value of the net returns shows that the proposed change in the farm plan will be profitable and should therefore be undertaken.

A partial budget answers to following questions as they relate to the proposed modifications in a farm plan (www.lec16.pdf): 1) What additional cost will be incurred? 2) What present income will be lost or reduced? The foregoing questions identify modifications which will reduce profit by either increasing costs or reducing returns. 3) What additional income will be received? and 4) What current costs will be reduced or lost? Answers to questions 3 and 4 identify factors which will increase profit by either generating additional income or lowering costs. Partial budget can take such forms as enterprise substitution, input substitution and size or scale of operation.

Enterprise substitution involves a total or partial substitution of one enterprise for the other. For example, if a farmer is growing a hectare of say maize but wants to change to a hectare of rice based on the assumption that both crops can be grown on the existing farm resources, the farm manager can then check all available price information and analyze the costs and returns for the alternative enterprises and use the resulting figures to generate a partial budget to enable the farmer to reach a decision.

Input substitution involves the substitution of an input for another. The total amount of input to be used is easily analyzed with a partial budget. In input substitution, the farmer is concerned about the financial implications of substituting a farm input for another. Examples include substituting family labour for hired labour, farm machinery for manual labour etc. Size or scale of operation as a form of partial budget involves changes in the size of the farm enterprise. It has to do with the cost implications and effectiveness of the expansion of the specified enterprise through buying or renting farm equipment and resources.

Complete Budgeting

Complete budgeting, also called total budgeting, is a tool for gathering and organizing data about an entire farm aimed at enabling decision-making about the management of farm resources. Complete budgeting considers the costs and returns of all crops, livestock, methods of production and aspects of marketing in a consolidated form. It is usually employed when contemplating complete overhaul of the organization and operation of an existing farm. Complete budgeting sets out to quantify all costs and returns and so present a complete picture of a farm. As the name implies, complete farm budgeting refers to the estimation of the costs and returns of all the enterprise combinations in a farm plan with a view to comparing the profitability and the feasibility of the alternative plans. Complete budgeting can be used together with partial budgeting in reaching decisions about modifications to be affected in the organization of a farm plan. However, differences exist between the two forms of budgeting.

Complete budgeting defines fundamental changes in the organization and operations of the farm, while partial budgeting focuses on addressing changes that are of a less fundamental nature. According to Welcome to Agriinfo.in (n.d.), other differences between complete budgeting and partial budgeting include: whole farm budget is considered as a unit while partial budget is adopted when minor aspects of farm organization is needed. In complete budgeting, both fixed and variable costs are calculated for working out the costs and returns. On the other hand, only the variable costs are considered in partial budgeting.

An example of complete budget in operation is the estimation of the costs and the returns of all the aspects of farms as a whole. Such aspects of the farms include the farm machinery items, crops, livestock, fisheries and poultry production, processing and marketing as a whole. Complete budget is required when radical changes are contemplated for the farm. An example of the operation of partial budget is the estimation of costs and returns of substituting one enterprise for another or estimating the costs and returns of hiring a machine as opposed to buying one.

Enterprise Budgeting

Enterprise budget is the estimation of the costs and returns and therefore the profitability of a specific enterprise. It is a listing of all the expected returns in both physical and value terms for a unit of a particular activity on the farm. Enterprise budget provides the best means to determine the expected profitability of a specific enterprise given the associated variable and fixed costs and returns. Farm decisions are easily and quickly made using the tool of enterprise budget.

CASH FLOW BUDGETING

A cash flow budget is a summary of the expected cash inflows and outflows for an enterprise over a given time period. Cash flow budget depicts the timing, frequency and volume of cash inflow and outflow for a specified time period for a named business. Cash flow budgeting is primarily employed as a forward planning tool, in estimating future borrowing needs and the loan repayment capacity of a particular business.

LINEAR PROGRAMMING

Meaning of Linear Programming

According to Harper and Lim (1982), linear programming is a term that covers a whole range of mathematical techniques that aim to optimize performance with respect to combination of resources. It is an optimizing technique in which all the variables are linear. According to Ferguson (n.d), a linear programming problem may be defined as the problem of maximizing or minimizing a linear function subject to linear constraints. In the views of Upton (1972), it is a systematic, mathematical procedure for finding the optimum or best plan for a given set of conditions. It is a planning tool which employs mathematical method of analysis, to establish the "best" or optimal combination of business activities to meet a clearly defined objective. For instance, as a farm planning technique linear planning can be used to determine the least-cost combination of crop and livestock products which will meet all the requirements of a balanced diet.

Linear Programming Assumptions

According to Harper and Lim (1982), before one can the linear programming technique, the following assumptions must hold: 1). There is a linear relationship between the output of each of the product and each of total quantity of each resource used, the total receipts, and the total costs, 2). Perfect divisibility of products and resources is assumed, 3). The basic principle of additivity is applicable throughout the calculations. For instance, if it requires N10 and N2000 to produce an egg and a full-grown chicken respectively then to produce a crate of 30 eggs and 12 live chickens will require (10x30)+(12x2000)= N24300, 4) . The problem involves only one function, for example the maximization of profit or the minimization of cost, 5). All external factors are unchanging, and 6). All quantities and values are known with certainty. In articulating the assumptions of linear programming technique, Upton (1972), presented three all-embracing and concise conditions in which the linear programming problem must be presented. They are: 1) a limited choice of several activities, 2) fixed constraints affecting the choice, and 3) straight line relations. Both Harper and Lim (1982), and Upton (1972), were pointing towards the same direction. However, the assumptions as presented by Upton (1972) are simpler, more concise and in terms of arrangement more compatible with the way a typical linear programming problem is generally arranged prior to solving it. The simple presentation is more easily understood by the average farmer who may which to adopt it as a farm planning tool. Let us take a closer look at the three components that are needed to solve a linear programming problem.

Limited Choice of Several Activities

The several activities may correspond with the alternative enterprises which can be produced in the farm business. On the other hand, the word 'activity' as used in linear programming obviously does not mean the same thing as the word 'enterprise'. Many different types of activities may be associated with the same enterprise where many alternative production methods are possible. For example, cassava cultivated in a mixture represents a different activity from cassava grown as a sole crop.

Fixed Constraints Affecting the Choice

With regard to the assumption of fixed constraint, this is aimed at restricting the combinations of activities which are feasible. Because the constraints are rigidly fixed, any plan that violates the constraints is to that extent considered not feasible. For example, if a linear programming problem has a labour constraint of say 40 days, a plan requiring 41 man-days is clearly not acceptable and is therefore not feasible. Note that 41 man-days is not feasible, the total labour requirement must be equal to or less than the stated constraint.

Straight Line Relations

The assumption of linearity means that irrespective of the number of a particular activity in a plan, the cost and returns stay same. The relationship between outputs and inputs are strictly assumed to be straight line. According to Upton (1982), assumption of linearity can give rise to unrealistic fractions. For example, if it requires 8 man-days to transplant 1 hectare of rice, then 16 man-days will be required to transplant 2 hectares of rice. Similarly, 4 man-days will be required to transplant 0.5 hectares of rice. When fractions occur especially when the problem involves machines or livestock, the fractions rounded up to the nearest whole number. The precision lost in the rounding process does not invalidate the solution.

Limitations of Linear Programming

The small-scale farmer who is desirous of employing the linear programming as a planning, it should be noted that manual calculation is somewhat mathematically very complex. Happily, available computers software simplifies the complex mathematical procedures and give efficient solutions. With or without the use of computer in the search for a solution, it should be fully appreciated that the linear programming technique has some obvious limitations. Some of these limitations have been made reference to in the text but cannot be overemphasized. Discussions under the limitations of linear programme benefited from the works of Lewis (2008) and Harper and Lim (1982).

The Basic Assumption That All Variables Are Linear Is Unrealistic

Linear programming treats all relationship as straight line. This assumption is incorrect, unrealistic and untrue in many real-life situations. Obviously, some variables possess some inherent randomness in real life situations.

Decisions Variables in Linear Programming Problem Is Meaningful If and Only If They Are Integer

Linear programming assumes that products are divisible into fractions and that resources can be employed in fractions of a unit. This assumption is not always the case and therefore cannot be justified. There are circumstances in which it is practically impossible to produce a fraction of a fish, poultry or livestock.. Where units can only be produced in whole numbers, a way round the problem of fractions when the linear programming technique is employed is to round up the fractions to the nearest whole number.

Other limitations of the linear programming technique include: Factors such as uncertainty, weather conditions etc. are not brought into the calculation; parameters are assumed to be constants this is not true in real life; linear programming procedure deals with only a single objective problem whereas in real life situations, there may be more than one objective; linear programming technique does not take into account time dimensions.

Standard Form of a Linear Programming Problem

Let us take a closer look at the standard form in which a typical linear programming problem is set up and with it further drive home the discussion. Here is a typical example.

$$\text{Maximize } Z = X_1 + X_2 \} \tag{1}$$

subject to the constraints

$$\left.\begin{array}{l} X_1 + 2X_2 < 4 \\ 4X_1 + 2X_2 < 12 \end{array}\right\} \tag{2}$$

and non-negativity requirements

$X_1 \geq 0, X_2 \geq 0\}$ (3)

The foregoing set of equations presents the standard form for setting up a linear programming problem. This consists of three parts namely (1) the function whose value is to be maximized. This is referred to as the objective function. If the objective function requires the total output (profit or total gross margin) to be maximized, it is a case of maximization problem and is called primal problem (Onimode & Osayimwese, 1984). The associated minimization problem is called the dual problem. The objective function usually contains at least two primary variables that are to be maximized (or minimized) subject to a series of (not less than two) linear inequalities called constraints. None of the variables may be negative. (2) refers to the ordinary structural constraints. The first inequality states that the demand for input X_1 for producing a good plus two times the demand for input X_2 to produce a good should be equal to or less than four. The second inequality requires that the total demand for four times input X_1 for producing a good plus the total demand for the production of two times X_2 for the production of a good should be equal to or less than twelve. (3) requires that the output of both goods X_1 and X_2 should be positive or zero.

Rules to Solve Linear Programming

When the linear programme problem is clearly stated, some basic rules are followed in finding a solution. According to Lewis (2008), some of the basic rules for solving a linear programming problem include:

Step 1: Mark the unknowns in the given linear programming by x and y.
Step 2: Formulate the objective function.
Step 3: Translate all the constraints in the form of equalities.
Step 4: Solve these equalities simultaneously.
Step 5: Find the values of x and y for which the objective function $z = ax + by$ has maximum or minimum value (as the case may be).

These steps are all applicable whether the problem is being solved by the computer or not.

Definition of Some Relevant Terms Related to Linear Programming

Before we go into solving a typical linear programming problem, let us define some relevant terms connected with the concept of linear programming as a planning tool. Some of the definitions were adapted from Lewis (2008).

Solution

A set of values $X_1, X_2, X_3, X_4, \ldots, X_n$ which satisfies the constraints of linear programming problem is referred to as its solution.

Feasible Solution

Any solution to a linear programming problem which satisfies the non negativity restrictions of the problem is called its feasible solution. Where a graph is used in solving the linear programming problem, the constraint lines act as production barriers. According to Harper and Lim (1982), the only practical final production mix must be the one that falls within all the barriers. This area is referred to as the area of feasible solution.

Optimal Solution

Any feasible solution which maximizes (or minimizes) the objective function of the linear programming problem is called its optimal solution. Any solution to a linear programming problem is said to be feasible, if none of any set of activities $Xj, j = 1, 2, 3, \ldots, n$, is negative. Note that if the linear programming problem is a maximization problem, then the set of values which satisfies the constraints is referred to as optimal solution to the primal problem. If the linear programming problem is a minimization problem, then the set of values which satisfies the constraints is referred to as optimal solution to the dual problem.

Slack Variable

Slack variable, also called a disposal activity is a variable introduced in an inequality to make it equality or an equation. Slack variable differs from the real activity because it is assumed there is no cost for not using land and the resource used up is the land itself (Upton, 1982).

Iteration

Iteration refers to the step by step trial and error procedure aimed at determining the optimal solution in a linear programming problem. According to Upton (1982), each step or iteration will yield a higher profit or gross margin than the previous one.

Theorem of Linear Programming

The basic theorem of linear programming is that in seeking the optimum solution to a linear programming problem, we need only to consider the results for the corner points of the feasible region. For a maximization problem of linear programming where a graph is used in finding the solution, the optimum solution is at the corner point of the feasible farthest from the origin. For a minimzation problem of linear programming where a graph is used in finding the solution, the optimum solution is at the corner point of the feasible region closest to the origin.

Other relevant terms in the concept of the linear programming technique as a planning tool are defined as they occur in the discussions in the text.

Examples of Linear Programming Problems

Let us take some examples linear programming problems and with them further drive home the discussion. Here is a simple example of a farm planning problem involving three alternative activities and four constraints. The objective is to maximize the total profit. The constraints are land, limited to 8 hectares, labour requirements in the months of May and November limited to 36 man-days and 32 man-days respectively and total labour limited to 320 man-days. The three activities consist of three alternative crops maize, cassava and yam. Each hectare of maize requires 8 man-days of May labour, no labour in November and 80 man-days of labour in total and yields a total profit of N10 per hectare. Each hectare of cassava requires 16-man-days in May, 20 man-days in November, 100 man-days in total and yields a total profit of N40 per hectare. Each hectare of yam requires 24 man-days in May, 25 man-days in November, 150 man-days in total and yields a total profit of

N50 per hectare. The question then is: what production plan maximizes the total profit? The problem can be presented as follows in Table 1.

In solving, we can proceed by presenting the problem in a standard linear programming form thus:

Let X_1, X_2, X_3 represent the number of units (that is hectare in this case) of the crops in the plan and Z to represent the total profit.

Maximize the total profit Z where

$$Z = 10X_1 + 40X_2 + 50X_3$$

Subject to the constraints

$$1\,X_1 + 1\,X_2 + 1\,X_3 \leq 8$$

$$8\,X_1 + 16\,X_2 + 24\,X_3 \leq 36$$

$$0\,X_1 + 20\,X_2 + 25\,X_3 \leq 32$$

$$80\,X_1 + 100\,X_2 + 150\,X_3 \leq 320$$

And the non-negativity requirements

$$X_1 \geq 0, X_2 \geq 0, X_3 \geq 0.$$

To solve the linear programming problem we must first transform the inequalities to equations. This is achieved by introducing slack variables to represent unused land and labour. For the unused land, the slack variable is:

Table 1. Basic data for a farm plan linear programming problem

	Activities			
	Constraint Level	**Maize**	**Cassava**	**Yam**
Total Profit		N10	N40	N50
Constraint				
Land (hectares)	8	1	1	1
May labour (man-days)	36	8	16	24
November labour (man-days)	32	0	20	25
Total labour (man-days)	320	80	100	150

Source: Adapted from Upton, 1982

$1\,X_1 + 1\,X_2 + 1\,X_3 + 1\,X_4 = 8$

For the non-use of May labour, we introduce the slack variable thus:

$8\,X_1 + 16\,X_2 + 24\,X_3 + 1\,X_5 = 36$

For the non-use of November labour, we introduce the slack variable thus

$0\,X_1 + 20\,X_2 + 25\,X_3 + X_6 = 32$

For the non-use of total labour, we introduce the slack variable thus

$80\,X_1 + 100\,X_2 + 150\,X_3 + X_{X7} = 320$

With these slack variables, we can rewrite the linear programming problem in an equation form:

Maximize $Z = 10X_1 + 40X_2 + 50X_3$

subject to the constraints

$$
\begin{aligned}
1\,X_1 + 1\,X_2 + 1\,X_3 + 1\,X_4 &= 8 \\
8\,X_1 + 16\,X_2 + 24\,X_3 + 1\,X_5 &= 36 \\
0\,X_1 + 20\,X_2 + 25\,X_3 + X_6 &= 32 \\
80\,X_1 + 100\,X_2 + 150\,X_3 + X_7 &= 320
\end{aligned}
$$

and the non-negativity requirements

$X_1 \geq 0, X_2 \geq 0, X_3 \geq 0, X4 \geq 0, X5 \geq 0, X6 \geq 0, X7 \geq 0.$

The slack variables have effectively changed the inequalities to equations. The non-negative requirements however remain as inequalities. According to Upton (1982), linear programming problem can be solved in three different ways namely intuitive approach, graphical approach and mathematical procedure. We can add the well-known fourth which is through the use of electronic computer.

The Intuitive Approach

The intuitive approach, although tedious is recommended for the average small-scale farmer who has chosen the linear programming technique as a planning tool. The recommendation is based on the fact that the approach is adaptable for solving the simplest of problems. Let us start this intuitive approach by looking at the feasible plan of leaving all the resources unused in disposal as set out in Table 2.

As shown in Table 2, the total profit in the plan is zero; so this is not an acceptable plan. We can therefore improve the plan to get the optimum solution. We start by choosing the activity that yields the highest total profit and expand it as far as possible. In this case, the chosen activity is the yam activity with a gross margin per unit of N50. We proceed by dividing each of the constraint levels by the corresponding requirement for the crop. There are 8 hectares of land of land available and each unit of yam requires 1 hectare; the land constraint permits a maximum of 8 units of yam. Each unit of yam needs 24 man-days of May labour out of a total of 36. This constraint will only allow 36/24 units to be produced. This is equal to 1.5units. November labour is similarly limited to 32/25 units of yam to be produced. This is equivalent to 1.28 units. The total labour supply restricts yam production to 320 /150 = 2.1 units. Obviously, the November labour (1.28 hectares) is the most limiting constraint for yam production. Any area of land larger than 1.28 hectares for yam is therefore not feasible.

According to Upton (1982), the introduction of 1.28 hectares of yam will correspondingly reduce the level of each of the other constraints. As a result, the land slack variable is reduced from 8 hectares to 6.72 hectares. In a similar way, the May labour slack variable is reduced by 1.28 times the requirement per hectare of 24 man-days. This is equivalent to 30.72 man-

Table 2. Beginning State of Linear Programming Problem

Real activities in the plan	None
Total profit	N0
Available land (hectares)	8
Available May labour (man-days)	36
Available November labour (man-days)	32
Available total labour (man-days)	320

Source: Adapted from Upton, 1982

days. This leaves a remainder of 5.28 man-days. The picture at the end of the first iteration is shown in Table 3.

We must now determine if it is possible to refine the plan as presented in Table 3. As can be seen, the critical constraint at this stage is the November labour. It is all in use and none is in disposal. To improve on this plan, we must select the enterprise which yields the highest profit per unit of this limiting resource. The choice is evidently from the remaining two enterprises of maize and cassava. The introduction of cassava competes with the available labour for yam in May. Introducing a unit of cassava requires 20 man days of November labour. Effectively, this means that 20 less must be employed on yam. Also, because yam production involves 25 man-days of November labour per unit, reducing the man-days by 20 man-days of labour will result in a reduction of 20/25 = 0.8 unit. This is interpreted as 1 hectare of cassava substitutes for 0.8 hectare of yam. The rate of technical substitution is 0.8. The total profit from a hectare of cassava is N40 and the opportunity cost (the alternative forgone) is the return from 0.8 hectare of yam is equal to 0.8 x N50 = N40. It is evident that the marginal net gain from removing November labour from yam to cassava enterprise is N40 – N40 = 0 and so nothing to be gained in the event of the change. A closer look shows that maize does not compete for labour in the critical November period; maize should therefore be introduced into the plan.

As we did in the first iteration, we now determine how far this activity can be expanded. We start by dividing the each constraint levels by the accompanying requirement for the maize enterprise based on the constraint levels as shown in Table 3:

Land (hectares) 6.7/1 = 6.7
May labour (man-days) 5.28/8 = 0.66

Table 3. State at end of initial iteration

Plan	1.28 hectares Yam
Total profit	N0 + (1.28 x N50) = N64
Available land (hectares)	8 – (1.28 x 1) = 6.7
Available May labour (man-days)	36 – (1.28 x 24) = 5.28
Available November labour (man-days)	32 – (1.28 x 25) = 0
Available total labour (man-days)	320- (1.28 x 150) = 128

Source: Adapted from Upton, 1982

November labour (man-days) No restriction
Total labour (man-days) 128/80 = 1.6

The critical constraint is now May labour. As a result, we can only introduce 0.66 hectare of maize and the constraint will be used up. The results at the end of the second iteration are as shown in Table 4.

As we again check if the plan can be improved, we discover that the May labour is now the critical constraint against further expansion of the plan. Next we check which activity gives the highest profit to the May labour. This is determined by dividing the total profit per unit by the May labour per unit. The following results as total profit per man-day for the May labour: N1.25 for the maize enterprise, N2.5 for the cassava enterprise, and N2.1 for the yam enterprise. Because the cassava enterprise offers the highest total profit to the May labour, the total profit can be increased by introducing cassava into the plan. The May labour requirement for cassava can be drawn from either maize or yam. Preferably, we withdraw it from maize because the return to May labour for maize is the lowest. On the other hand, November labour must necessarily be drawn from yam production because maize does not use any labour in this month of November. As stated earlier, the rate of technical substitution of cassava for yam with respect to the November labour is 0.8. But at this stage of the iteration, transfer of December labour from yam to cassava will release June labour and eventually allow the maize yield to be increased.

So, as argued by Upton (1982), if a unit of cassava that needs 16 man-days of May labour is introduced at the expense of 0.8 unit of yam which requires 0.8 x 24 = 19.2 man-days, the May labour requirements i. effectively reduced by 19.2-16 = 3.2 man-days. So, each unit of cassava introduced releases 3.2 man-days of May labour. Note that maize gives the least return to the May labour; however it does not require any November labour. So, maize can be

Table 4. State at end of the second iteration

Plan	1.28 hectares Yam + 0.66 hectare of Maize
Total profit	N64 + (0.66 x N10) = N70.6
Available land (hectares)	6.7 – (0.66 x 1) = 6.04
Available May labour (man-days)	5.28 - (0.66 x 8) = 0
Available November labour (man-days)	0 – (0.66 x 0) = 0
Available total labour (man-days)	128 – (0.66 x 80) = 75.2

Source: Adapted from Upton, 1982

expanded to use the extra May labour without competing with the either cassava activity or the yam activity. For each 3.2 man-days of May labour, 3.2/8 = 0.4 unit of maize to be introduced. As can be seen, the introduction of a unit of cassava results in a decrease of 0.8 units of yam and an increase of 0.4 unit of maize. So the return is N40 the cassava added to 0.4 x N10 = N4 as compared with an opportunity cost of 0.8 x N50 = N40. Because there is a marginal net gain in transferring the November labour from the yam enterprise to the cassava enterprise, it is therefore profitable to make the change. Obviously, the marginal net gain from moving the November labour from the yam enterprise to the cassava enterprise is N44 – N40 = N4.

Before we start the next iteration to determine how far this plan can be expanded by dividing the constraint levels with the relevant requirement for the cassava enterprise, it should be noted that the area of yam is now a constraint on the amount of cassava to be introduced. This is so because cassava is substituting for yam. There are 1.28 hectares of yam in the plan and each hectare of cassava substitutes for 0.8 hectare of yam; the total area of cassava which can be introduced is 1.28/0.8 = 1.6 hectares; this will allow the expansion of maize by 1.6 x 0.66 = 1.06 hectares. Because there is surplus of the constraints of land and total labour, the change can be effected. Yam activity is eliminated from the plan.

Table 5 shows the state of the iteration at the end of the third iteration.

We repeat the process of determining the possibility of improving the plan as shown in Table 5. In this intuitive approach to solving a linear programming problem, Upton (1982) observed that if yam is reintroduced, each unit would compete with cassava for November labour and with both cassava and maize for June labour. The introduction of 1 unit of yam would require a reduction of 25/20 = 1.25 units of cassava to release the required December labour.

Table 5. State at end of third iteration

Plan	1.6 hectares cassava and 0.4+ 0.66 hectare of Maize = 1.1 hectare maize (1.28 hectares of yam are out)
Total profit	70.6 – (1.28 x 50) + (1.6 x 40) + (0.66 x 10) = 77.2
Available land (hectares)	6.04 + (1.28 x 1) - (1.6 x 1) - (0.66 x 1) = 2.5
Available May labour (man-days)	0 + (1.28 x 24) – (1.6 x 16) – (0.66 x 8) = 0
Available November labour (man-days)	0 + (1.28 x 25) – (1.6 x 20) – (0.66 x 0) = 0
Available total labour (man-days)	75.2 + (1.28 x 150) - (1.6 x 100) - (0.66 x 80) = 54.4

Source: Adapted from Upton, 1982

A unit of yam will need a decrease of 25/20 = 1.25 units of cassava to release November labour. It will also release 1.25 x 16 = 20 man-days of May labour. However, 1 unit of yam needs 24 man-days in the month of May. That is 4 man-days extra. Therefore, the maize enterprise has to be reduced by 4/8 = 0.5 unit to release May labour. The total cost forgone for introducing 1 hectare of cotton is 1.25 x N40 + 0.5 x 10 = N55. Because the total profit from 1 hectare of yam is N50, this therefore is not profitable.

The expansion of the cassava enterprise is not possible because all the existing supply of November labour is needed for the present area of the cassava enterprise.

With respect to the expansion of maize enterprise, it can be clearly seen that it would compete with cassava for labour in May for 1 unit of Maize requires 8 man-days in the month of May. Therefore, it would substitute for 8/16 = 0.5 unit of cassava. The opportunity cost is 0.5 x N40 = N20 as return 1 hectare of maize of only N10. It can therefore be concluded that it is not possible to increase the total profit and so the plan in Table 5 cannot be profitably expanded further. The plan in Table 5 is therefore the optimal solution.

We have thus far highlighted the intuitive approach as a method of solving a linear programming problem. Upton (1982) described the intuitive approach as somewhat tedious and can only be used for the very simplest of linear programming problems. The approach however illustrates the reasoning behind solving a linear programming problem.

Evidently, linear programming is simply mathematically very complex. According to Harper and Lim (1982), to avoid such complexity in the actual solving of problems, mathematicians have reduced the techniques to a series of simple but lengthy procedures. Harper and Lim (1982) also described the "how" of solving linear programming problems as opposed to the "why" of the procedure. This book has also adopted the procedure of describing the "how" of solving a linear programming problem. Let us continue with another approach to solving a linear programming problem. The other approach is called the graphical approach.

The Graphical Approach

The graphical approach easily lends itself to be followed for application by small-scale farmers with an eye on engaging in agriculture as a business. According to Olayide and Heady (1982), Upton (1982), and Harper and Lim

(1982), the graphical approach is useful in solving two-dimensional linear programming problems. Harper and Lim (1982), further added that there should be two or more constraints so that it is not immediately clear what enterprise mix will maximize the profit or, as the case may be, minimize the cost. In the views of Arene (2008), the graphical approach to solving a linear programming problem involves two steps: the graphical determination of the region of feasible solution; and the graphical determination of the objective function.

Let us begin by reducing the linear programming problem we solved using the intuitive approach a two-activity problem and then demonstrate how it can be solved graphically.

The problem is therefore:

In solving, we can proceed by presenting the problem in a standard linear programming form thus:

Let X_1, X_2, represent the number of units (that is hectare in this case) of the crops in the plan and Z to represent the total profit.

Maximize the total profit Z where

$$Z = 10X_1 + 40X_2$$

Subject to the constraints

$$1\,X_1 + 1\,X_2 \leq 8$$

$$8\,X_1 + 16\,X_2 \leq 36$$

$$0\,X_1 + 20\,X_2 \leq 32$$

$$80\,X_1 + 100\,X_2 \leq 320$$

And the non-negativity requirements

$$X_1 \geq 0, X_2 \geq 0.$$

In determining the non-negativity requirements of the problem, we start the search from the positive quadrant of the x and y axes of a typical graph.

We then plot the possible combinations of cassava and maize on the graph. Let us also state the inequalities as equalities thus:

$1\,X_1 + 1\,X_2 = 8$

$8\,X_1 + 16\,X_2 = 36$

$0\,X_1 + 20\,X_2 = 32$

$80\,X_1 + 100\,X_2 = 320$

Next we set the boundaries for the units of maize in the graph in Figure 1. This is marked by a straight line drawn from the *x*-axis to the *y*-axis. The respective values at the *x* and *y* axes are obtained from the different equations as follows: For the equation $1\,X_1 + 1\,X_2 = 8$, to operate under this land constraint we have to produce either 8 hectares of maize or 8 hectares cassava. Intermediate combinations of the two enterprises are also possible without violation of the stated land constraint. We then plot these values in the relevant axes on the graph as represented by line AB. For the equation $8\,X_1 + 16\,X_{2} =$ 36, which represents the May labour constraint, if all this resource is used up in maize production, 36/8 = 4.5 would be feasible. If all this resource is used up in cassava production, 36/16 = 2.3 would be feasible. We then plot these values in the relevant axes on the graph as represented by line CD. For the equation $0\,X_1 + 20\,X_{2} =$ 32, note that the maize enterprise, does not require labour in November. The value of this constraint as represented rises vertically from 32/20= 1.6 units on the x axis, line EFX. With respect to total labour as a constraint, if all this resource is used up on maize, X_1, production 320/80 = 4 units will be feasible. If all this resource is used up on cassava, X_2, production 320/100 = 3.2 units will be feasible. We then plot these values in the relevant axes on the graph as represented by line GH.

The region of feasible solution is set by the area GIFE in which all the technical and non-negativity constraints are satisfied. Next we determine the maximum possible profit that can be obtained from the combination of the maize and cassava enterprises. This is done by constructing an iso-profit line, broken line CD on the graph, that is tangent to the farthest point right of the frontier of the region of feasible solution. Iso-profit line is a line drawn joining points of equal profits on a graph. In this case, it is a line joining all combinations of units of maize and units of cassava which yield a given amount of total profit. Lower iso-profit lines are evidently not considered. The values of maize, $X_{1,}$ and cassava, $X_{2,}$ are read off and substituted in the equation $Z = 10X_1 + 40X_2$ to determine the monetary value of the maximum total profit that can be obtained at the point of tangency. As shown in Figure 2, at the point in which the line (broken line CD) touches the farthest point on the feasible, X_1 has the value of 1.1 and X_2 has the value of 1.6. Substituting these vales into the profit function we have:

$Z = 10(1.1) + 40\ (1.6) = 75$. It is evident that the maximum total profit that can be obtained is N75.

We have just shown how to solve a maximization problem of linear programming. It should be noted the problem can also be concerned with a minimsation problem requiring the determination of factors to minimize the cost function. The procedure is the same as that for the graphical solution of a maximisation problem. A linear programming problem requiring minimization of cost as the objective function differs from that of maximization in terms of the position of the point of the feasible region. As earlier stated under the definition of terms related to linear programming, a maximization problem of linear programming where a graph is used in finding the solution, the optimum solution is at the corner point of the feasible farthest from the origin. For a minimzation problem of linear programming where a graph is used in finding the solution, the optimum solution is at the corner point of the feasible region closest to the origin. So, for a minimization problem, after graphing the relevant values and clearly marking out the feasible region, the line to be drawn tangent to the lowest point on the feasible region should be closest to the origin and the relevant values read off as usual and substituted in the objective function.

As stated earlier linear programming problem can be solved in three different ways namely intuitive approach, graphical approach and mathematical procedure. We have highlighted the intuitive approach and the graphical approach. Next is the mathematical approach.

The Mathematical Approach

The mathematical approach also called the simplex method. Linear programming problems with more than two variables cannot be solved using the graphical method, so the simplex method is used. According to Harper

Figure 2. Graphical solution of maximization linear programming problem

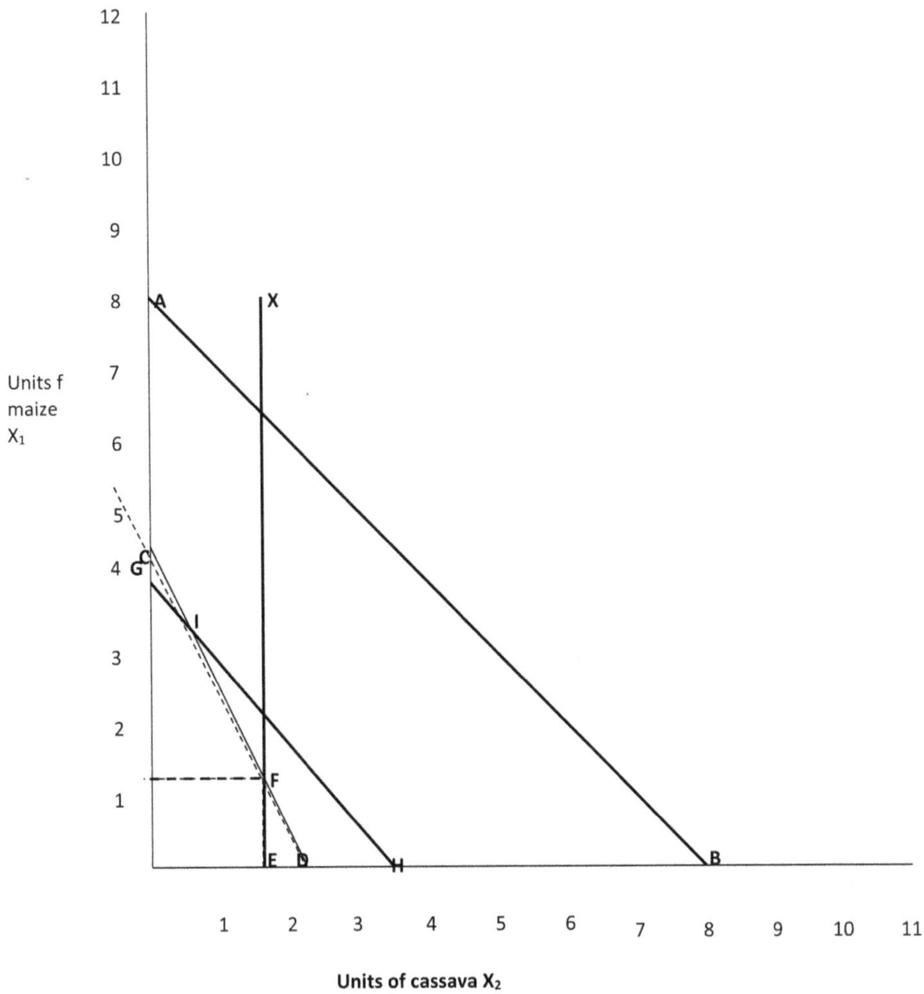

and Lim (1982), the simplex technique is used where there are two or more products (enterprises); and where there are two or more constraints operating so that it is not immediately apparent which product or enterprise mix will maximize the profit or minimize the cost. According to Olayide and Heady (1982), the method is based on the property that the optimum solution to the linear programming problem, if the solution exists at all, can be found in one of the basic feasible solutions. In the simplex method, the initial step is to determine a basic feasible solution. This involves obtaining a set of basis vectors and a set of non-basis vectors. Each variable corresponds to a column in the tableau. A tableau is another name for matrix. Matrix is a rectangular arrangement of numbers. If a column in a tableau is cleared out and so has only one non-zero element in it, then that variable is a basic variable. If a column is not cleared out and has more than one non-zero element in it, that variable is non-basic and the value of that variable is zero. (Linear Programming: Simplex Method, n.d.).

The simplex method is employed for linear programming problems with many variables; such symbols as x_1, x_2, x_3 and x_4 are used in labeling the variables; all constraints must be expressed in the form $a_1x_1 + a_2x_2 + a_3x_3 + \ldots \leq b$

The steps in using the simplex method in solving a linear programming basically involve the following steps (Johnson, n.d.).

1. The first step is to convert each constraint (a linear inequality) into a linear equation. This is done by adding a nonnegative variable, called a slack variable to each constraint. The slack variables are use in converting each constraint into a linear equation.
2. Introduce slack variables as necessary and then write the initial simplex tableau for the linear programming problem.

Maximize: $Z = 5x_1 + 3x_2 + 7x_3$

Subject to: $4x_1 + 3x_2 + 2x_3 \leq 60$

$3x_1 + 4x_2 + x_3 \leq 24$

$X_1 \geq 0, x_2 \geq 0, x_3 \geq 0$

Solution: The Slack Variables are x_4 and x_5

$4x_1 + 3x_2 + 2x_4 = 60$

$3x_1 + 4x_2 + x_3 + x_5 = 24$

The objective function $Z = 5x_1 + 3x_2 + 7x_3$ needs to be rewritten so that all the variables are on the left-hand side: $-5x_1 - 3x_2 - 7x_3 \geq 0$.

Next step is to set up the initial tableau. The linear programming problem is written in matrix form called the simplex tableau. The last row of the simplex tableau contains the objective function with all variables on the left-hand side.

$$\text{The initial simplex tableau is }\begin{matrix} X_1 & X_2 & X_3 & X_4 & X_5 & Z \end{matrix}\\ \left\{\begin{array}{cccccc|c} 4 & 3 & 2 & 1 & 0 & 0 & 60 \\ 3 & 4 & 1 & 0 & 1 & 0 & 24 \\ \hline 5 & -3 & -7 & 0 & 0 & 10 & \end{array}\right\}$$

Except for the entries 1 and 0 on the right end-the numbers in the bottom row of a simplex tableau are called indicators. The indicators in this tableau are -5 -3 -7 0 and 0. The column containing the most negative indicator is called the pivot column.

Next step in solving a linear programming problem using the simplex method is to locate the most negative indicator. This indicator determines the pivot column. In the simplex tableau shown as an example, the third column is the pivot column, since -7 is the most negative indicator.

Next step is to use the positive entries in the pivot column to form the necessary quotients for determining the pivot. If there are no positive entries, no maximum solution exists. If two quotients are equally the smallest, let either determine the pivot.

Next step is to divide the pivot row by the pivot to change the pivot to 1. Then use row operations to change all other entries in the pivot column to 0. This is achieved thus: for each positive entry in the pivot column, divide the number in the far-right column of the same row by the positive number in the pivot column. The row with the smallest quotient is called the pivot row. The entry in the pivot row and pivot column is the pivot. The pivot is the number in the matrix chosen to change to a 1 and all other entries in that column changed to zeros. The smallest quotient is 24 (60/2 = 30 and 24/1 =24), so the pivot is 4. Pivoting is complete once all the indicators are nonnegative.

The foregoing step by step description of the simplex method of linear programming can be summarized as: Convert each constraint to an equation by adding slack variables; set up the initial tableau; locate the most negative indicator. This indicator determines the pivot column; use the positive entries in the pivot column to form the necessary quotients for determining the pivot. If there are no positive entries, no maximum solution exists. If two quotients are equally the smallest, let either determine the pivot; divide the pivot row by the pivot to change the pivot to 1. Then use row operations to change all other entries in the pivot column to 0. If the indicators are all positive or 0, you have found the final tableau; set each non-basic variable equal to 0 and solve the system for the basic variables. The maximum value is the number in the lower right-hand corner of the tableau

(Johnson, n.d.). The interested reader should refer to the references stated in this section for more information on the concept of linear programming.

The question may arise: how is the average farmer, who is relatively poor and barely literate (Opara, 2010), apply such budgeting techniques as described in the chapter to his/her farm operation? This chapter in particular and the entire book in general brings to the reading and the farming public some best practices that if employed go a long way in making agriculture as a going concern. We should not continue doing a thing the same way all over and expect to get a solution to an existing problem. Essentially, the role of extension service and literacy programme targeted at the farmers and the setting up of support mechanism to help in the dissemination of the budgeting techniques and processes should be explored.

SOLUTIONS AND RECOMMENDATIONS

This chapter focuses on the basic ways of planning an investment using the instrument of budget. There are some tools employed in the farm planning process. The two basic planning tools available are budgeting and linear programming. Enterprise budgets show which parts of the farm operation are making a profit and which parts are not. Whole-farm budgets are best at analyzing the complementarity among different parts of a farm. In both cases, reliable and valid record keeping is essential in keeping track of the progress made by an enterprise in achieving the set objectives.

Farm planning with the associated use of the budgeting tool is successful to the extent of the availability, validity and reliability of the records that go into them. It cannot be overemphasized that for agricultural enterprises to be run as a business there is the urgent need for the farmers to keep farm records especially in view of their special role in the process of farm planning

For agriculture to be practiced as a business, it is recommended that the farmers should essentially base their entrepreneurial activities on the profitability analysis borne out the tools of farm planning. This is a direct departure from unplanned and uncoordinated committing of scarce resources to an enterprise without tracking of the proportionate returns resulting from such investment.

It is also recommended that formal and informal farmer training programmes and extension services together with the setting up of support mechanism to help in the dissemination of the budgeting techniques and processes should be explored.

FUTURE RESEARCH DIRECTIONS

Small-scale farmers have over the years been known not to keep valid and accurate farm records. However, it is not empirically known to what extent this lack of record keeping can be attributed to poor profitability of the farm enterprise. Future research is suggested to focus this area.

CONCLUSION

As agricultural enterprise becomes more business oriented, and environmental regulations become more site-specific, farm planning becomes more useful. It is against this backdrop that present and prospective farmers should understand the importance of farming planning and its relevance in the transformation of the agricultural sector to a new height of profitable venture. Understanding the importance of farm planning and applying the understanding in agriculture as a business will enable the farmer achieve profitability.

REFERENCES

Arene, C. J. (2008). *Agricultural Economics: A functional approach*. Enugu, Nigeria: Prize Publishers.

Blasingame, J. (2011). *The difference between business plan and planning*. Retrieved October 29, 2017 from http://www.forbes.com/sites/jimblasingame/2011/06/10/difference-between-plan-planning/#38366646a253a

Department of Trade and Economic Development. (2010). *Business Facts*. Government of South Australia: Department of Trade and Economic Development. Retrieved from www.southaustralia.biz

Dueling, T. N., & Sherill, W. W. (2006). *Entrepreneurism: Exploring entrepreneurism from a business process perspective*. Atomic Dog Publishing.

Federal Ministry of Agriculture and Rural Development (FMARD). (2016). *The Agriculture promotion policy (2016 – 2020) building on the successes of the ATA, Closing key gaps policy and strategy document*. Abuja, Nigeria: Federal Ministry of Agriculture and Rural Development.

Ferguson, T. S. (n.d.). *Linear programming: A concise introduction.* Retrieved August, 2017, from https://www.math.ucla.edu/~tom/LP.pdf http://dmc122011.delmar.edu/math/pjohnson/Webpage/businessmath/notes/7.4.pdf

Harper, W. M., & Lim, H. C. (1982). *Operational Research.* The English Book Language Book Society and Macdonald & Evans London.

Janke, R. R. (2000). *Whole-Farm Planning for Economic and Environmental Sustainability.* Kansas State University.

Kuratko, D. F., & Hodgetts, R. M. (1998). *Entrepreneurship: a Contemporary Approach* (4th ed.). New York: Harcourt- Brace College Publishers.

Lewis, C. (2008). *Linear Programming: Theory and Applications.* Retrieved August, 2017 from https://www.whitman.edu/Documents/Academics/Mathematics/lewis.pdf

Marzec, E. (n.d.). *Business plan vs. feasibility study.* Retrieved October 28, 2017 from http://smallbusiness.chron.com/business-plan-vs-feasibility-study-43382.html

Mesly, O. (2017). *Project feasibility- Tools for uncovering points of vulnerability.* New York: Taylor and Francis.

Minnesota Institute for Sustainable Agriculture (MISA). (2011). *Introduction to whole farm planning combining family, profit, and environment.* Minnesota Institute for Sustainable Agriculture. Retrieved from www.misa.umn.edu

National Universities Commission (NUC). (n.d.). *Entrepreneurship studies for distance learners in the Nigerian university system.* Retrieved October 28, 2017 from www.freebookcenntre.net

Olayide, S. O., & Heady, E. O. (1982). *Introduction to agricultural production economics.* Ibadan, Nigeria: Ibadan University Press.

Onimode, B., & Osayimwese, Iz. (1984). *Mathematics for economics and business*. George Allen & Unwin London.

Opara, U. N. (2010). Personal and Socio-economic Determinants of Agricultural Information Use by Farmers in the Agricultural Development Program Zones of Imo State, Nigeria. *Library Philosophy and Practice*, *434*. Available: http://unllib.unl.edu/LPP/opara.htm

Thompson, A. (2005). *Entrepreneurship and Business Innovation: The Art of Successful Business Startups and Business Planning*. Entrepreneurship and Business.

Touchie, R. D. (2005). *Preparing a Successful Business Plan*. EPP Book Services.

Women in Business. (2010). Retrieved from www.womeninbusiness.about.com/od/busnessplan/a/feasibilitystud

Linear Programming, Simplex Method. (n.d.). Retrieved November 16, 2016, from https://people.richland.edu/james/ictcm/2006/simplex.html

Staff, I. (2015, August 11). *Feasibility Study*. Retrieved November 16, 2016, from https://www.investopedia.com/terms/f/feasibility-study.asp

Department of Economic Development, Jobs, Transport and Resources. (2017, June 8). *Farm management*. Retrieved June 15, 2017, from http://agriculture.vic.gov.au/agriculture/farm-management

Welcome to Agriinfo.in. (n.d.). Retrieved November 16, 2016, from http://www.agriinfo.in/

Johnson, P. (n.d.). *Linear Programming: The Simplex Method*. Retrieved November 16, 2016, from www.delmar.edu/math/pjohnson/Webpage/businessmath/notes/7.4.pdf

Chapter 10
Cost of Capital and Methods of Charging Interest

ABSTRACT

In the course of running a business, debt and equity are employed in the financing decision. The employment of debt and equity in the financing mix of an enterprise involves costs. The investing farmer should keep track of the cost of capital invested in an enterprise if only to ensure that they do not exceed the returns. It is against this background that this chapter is devoted to the cost of capital ad methods of charging interest. It is concluded that it makes business sense to equip farmers with the knowledge that interest rate on loans is subject to negotiation and that the amount paid as interest on a loan depends on the method adopted in calculating the interest rate. It is recommended that in employing funds for running an enterprise, the farmer should be committed to the obligations of due repayment and also make profits from the borrowed funds.

INTRODUCTION

In the course of running a business, capital items and funds are procured and invested in the business at a cost and for expected return. The cost includes both that of procurement and replacement. The purpose of this chapter is to bring to the investing farmer relevant discussions on the cost of capital and methods of charging interest rate. The chapter focuses on the costs associated with borrowing and the procedures for calculating interest charges. It also attempts to differentiate between cost of capital and cost of funds.

DOI: 10.4018/978-1-5225-3059-6.ch010

The rest of the chapter is organized as follows: Cost of Capital and its Calculation including Calculation of Debt, Cost of Equity including Calculation of Equity and Calculation of Weighted Cost of Debt, Cost of Funds, Interest Rate, Computation of Interest Rate, Simple Interest Rate, Compound Interest, Interest Rate and Agricultural Productivity in Nigeria. The chapter concludes with recommendations.

Discussions in this chapter benefitted from the ideas obtained from such sources as the following:

- http://www.investopedia.com/terms/i/interestexpense.asp#ixzz3vgm16PHW,
- http://www.investopedia.com/terms/c/costoffunds.asp#ixzz3qDRrGJ63,
- http://www.investopedia.com/terms/i/interestrate.asp#ixzz3qDVApaRQ,
- http://www.investopedia.com/terms/c/costofequity.asp#ixzz3qDXJNFDK,
- http://www.investopedia.com/terms/r/rateofreturn.asp#ixzz3qDdJJlBn,
- http://www.investopedia.com/terms/c/costofdebt.asp#ixzz3qDe2EyYv,
- https://www.investopedia.com/terms/w/wacc.asp#ixzz3tdarmPdM

COST OF CAPITAL

Cost of capital refers to the cost of funds employed in a business. It can be construed as the financial involvement in a named capital investment in a business for a given period. Cost of capital is composed of the costs of the various components of financing an enterprise. In relation to capital budgeting, it represents the expected minimum rate of return on an investment required by the investor in order to compensate for the associated business and financial risks.

According to Maharshi Dayanand University (2004), cost of capital, often called the minimum required rate of return (Anon, 2014), is the rate that should be earned to satisfy the required rate of return of a firm's investors. It can also be viewed as the rate of return on investments at which the price of a firm's equity share will remain unchanged. This minimum required rate of return threshold must be crossed before the enterprise can generate profit and so be attractive to investors. Cost of capital can be made up of the cost of equity and the cost of debt. It can also be made up of either the cost of equity or the cost of debt. If the enterprise in question is solely financed through equity, then the cost of capital is equal to the cost of equity. It is the cost of debt if the business is financed solely through debt. Therefore, the cost of capital is a reflection of the mode of the financing mix of a named enterprise.

Calculation of Cost of Capital

The cost of capital is computed as a weighted average of the various capital components (Anon, 2014). It is a common practice for many enterprises to use a combination of equity and debt to finance their activities, hence, the cost of capital is estimated from the weighted average of the cost of equity and the cost of debt; this is widely known as the weighted average cost of capital (WACC). According to Anon (2014), the cost of capital is computed as a weighted average of the various capital components, which include debt, preferred stock, common stock, and retained earnings. However, not all businesses have this capital structure configuration. For instance, the capital structure involving the capital components of preferred stock and common stock is somewhat beyond the range of operation of the small-scale farmer. Discussions on them are also not included here. Lee, Boehlje, Nelson, and Murray (1980), views on computation of weighted average cost of capital are more relevant to the focus of this book which is on small scale farmers. According to Lee, Boehlje, Nelson, and Murray (1980), weighted average cost of capital involves estimating the after-tax cost of borrowed capital and the after-tax cost of equity capital and then computing an overall capital structure of the enterprise. Let us look at the ways the cost of debt and the cost of capital are calculated. After focusing on these cost components, the two will be brought together as a composite cost of capital.

The formula for the calculation of the weighted average cost of capital is:

$$K_o = w_d K_d + w_e K_e$$

Where Ko = weighted average cost of capital

Kd = cost of debt capital

Ke = cost of equity capital

w_d and w_e = proportions of debt and equity capital respectively in the desired or optimal capital structure of the enterprise

Calculation of Cost of Debt

The before-tax cost of debt capital refers to the contractual rate of interest on loans including administrative and sundry charges (Staff, 2016). The cost of debt is derived by adding a default spread to the risk-free rate; the magnitude of the spread depends on the credit risk of the enterprise in question. The cost of debt refers to the interest rate paid by the borrower on the debt in question. Since interest expense is tax-deductible, the after-tax cost of debt has to be calculated.

According to Lee, Boehlje, Nelson, and Murray (1980), the after-tax cost of debt is obtained using the formula:

$$K_d = i\,(1\text{-}t) + R$$

where K_d = cost of debt financing

i= before-tax interest rate on loans

t= marginal income tax rate

R= internal credit rationing premium for financial risk and credit loss

Marginal tax rate is the amount of tax paid on an additional amount of income. The marginal tax rate increases as the income increases. The R in the formula can also be interpreted as the noninterest cost of borrowing. Because the use of debt financing is associated with increased financial risk and decreased credit reserves, an internal credit rationing premium is added to the interest rate. The value assigned to the internal credit rationing premium is at the discretion of the investor or the farmer doing the analysis. The cost of debt so derived is then used as minimum cost for all debt.

COST OF EQUITY

The cost of equity can be defined as the minimum rate of return that an enterprise must generate on the equity-financed component of its investment portfolio in order to leave unchanged the market price of its stock (Staff, 2015a). In financial theory, cost of equity is the return that stockholders require for a firm. Cost of equity represents the compensation that investors demand in exchange for owning the asset and bearing the risk of ownership.

Since equity-holders expect to receive a certain return on their investment in a firm, the equity holders' required rate of return is a cost to the firm; if the firm fails to earn the expected return, the shareholders will tend to disinvest with negative consequences to the market value of the firm. To all intents and purposes, cost of equity is the amount that a firm must spend in order to maintain a share price that is satisfactory to the firm's investors.

Calculating Cost of Equity

One of the ways of measuring cost of capital is through the use of the dividend capitalization model. This method approximates a future dividend stream based on a company's dividend history and an assumed growth rate, and computes the market capitalization rate that equates it with the current market price. In the case of such enterprises as sole proprietorships and partnerships that do usually not ordinarily distribute profits as dividends, the enterprise's dividend-paying capacity is estimated from its average net income and average cash flow and compared with the dividends actually paid by a similar size enterprise. The formula for calculating cost of equity under the dividend capitalization model is shown in Figure 1.

For example, if a company's expected dividend per share for next year is N3.00, the current market value of the stock is N60.00; earnings and dividends per share are expected to grow about 5% per annum, the company's cost of equity capital is:

Figure 1. Formula for calculating cost of equity

$$\text{Cost of Equity} = \frac{\text{Dividends per Share (for next year)}}{\text{Current Market Value of Stock}} + \text{Growth Rate of Dividends}$$

$K_e = 3/60 + .05 = 10\%$

The critical factor here is actually and realistically measuring the growth in dividends per share from the perspective of the investors. The expected growth in dividends is not directly observable; it must therefore be estimated. This is the crux of the difficulty in measuring the cost of equity capital. Related to this difficulty is the fact that the rate of return demanded by equity investors is not as clearly defined as it is by lenders.

Generally, the procedures for the estimation of cost of capital are strictly applicable to business entities whose shares are publicly traded. The procedure adopted here is obviously related to one of the models of estimating the cost of equity capital raised by issuing new shares in the capital market. Because most farm enterprises shares are not publicly traded, it is germane that we extend the discussion to some guidelines for estimating the cost of capital for a typical small-scale enterprise.

Lee, Boehlje, Nelson, and Murray, (1980), presented the following guidelines for estimating the cost of equity capital for small-scale enterprises. Evidently, equity capital is not free; it is not without cost. As a result, the cost of equity should be equal to or greater than that of debt. For one thing, surplus funds in a business enterprise can be used to reduce the outstanding debt obligations and subsequently the interest expense. For another, the suppliers of capital assume more than the creditors. By how much the cost of equity will be greater than the cost of debt is clearly a personal consideration.

Some questions are pertinent in determining by how much the cost of equity capital will be more than the cost equity for small-scale enterprises namely: What annual rate of return on equity after taxes has the enterprise been earning in the past years? If the assets of the enterprise are sold, all debts repaid, and the resulting networth invested in nonfarm assets that have similar risk, what rate of return would I earn? Lee, Boehlje, Nelson, and Murray (1980), concluded that the estimate for the cost of equity capital can be based on past rates of return, on the opportunity of investment on similar enterprises or on a combination of the two approaches while ensuring that the cost of equity essentially exceeds the cost of debt.

Calculating Weighted Cost of Capital

The cost of capital for an enterprise is based on the weighted average of the costs of debt and that of equity (Staff, 2015b). To calculate a firm's weighted cost of capital, it is needful to first determine the costs of such component financing sources as cost of debt, cost of preference capital and cost of equity (Fernandes, 2014). For our purposes in this book, the individual financing sources are the debt and the equity. The weights to the cost are the debt to equity ratio in the desired capital structure of the enterprise. Although the determination of the optimal or desired capital structure for the enterprise is mostly subjective, it can be arrived at by asking a sample of farmers or inventors in a similar enterprise within the same location what could be considered desirable or optimal debt to equity ratio.

According to (Anon, 2014), an enterprise's cost of capital is the weighted average of the individual capital costs, with the weights being the proportions of each type of capital used. It is the summation of the percentage of the costs of all capital components multiplied by the cost of capital for each source. The computation of overall cost of capital is illustrated in the following simplified example. It is simplified to the extent that component financing sources are the debt and the equity only. Assume that the capital structure of the proportions of financing that a farmer intends to use over time is:

Cost
Debt N30,000 15%
Equity N70,000 15%
Total N100,000

These proportions are applied to the assumed after-tax costs (Table 1):

The overall cost of capital is 9.6%. This means that 9.6% is a measure the cost the firm's cost of capital. It represents the minimum rate of return that the firm must earn and so becomes the reference point, the benchmark for evaluating the profitability of investments for enterprise expansion.

Table 1.

Source	Weights Cost	Weighted Cost
Debt	30%(a) 11%	3.3% (b)
Equity	70% 9%	6.3%
	100%	9.6%

a= 30,000/100,000 x 100= 30%; b= 30% x 11%= 3.3%

COST OF FUNDS

A concept related to cost of capital is that of cost of funds. Cost of funds is the interest rate paid by financial institutions for the funds that they deploy in their business (Staff, 2015c). Financial institutions (FIs) refer to establishments that deal with such financial transactions as investments, loans and deposits. For lenders such as banks and other financial institutions, cost of funds is determined by the interest rate paid to depositors on financial products including savings accounts and current account deposit. For borrowers, cost of funds is determined by the costs they must incur to gain access to funds.

In typical credit transactions, the interest rate a lender charges essentially covers cost of funds, risk premium, and loan administration costs. The cost of obtaining funds depends on the source of the fund and on the general economic condition. Lenders encounter two kinds of risks namely default risk and market risk. Default risk is the possibility that the amount borrowed will not be repaid as stipulated in the contractual loan agreement. Market risk, also known as interest rate risk refers to the tendency of changes in the general level of interest rates during the course of the loan.

Cost of funds and net interest spread enable banks make money. Banks make money through the interest rates they charge on loans as well as debt securities they own and other equity products that consumers, companies and large-scale institutions need. The interest rate banks charge on loans granted to customers must be greater than that of the interest rate that the banks pay for the use of funds.

Before we attempt to differentiate between cost of capital and interest rate, let us first of all look at the concept of interest rate in its proper context.

INTEREST RATE

According to Drummond and Godwin (2004), interest rate is the price of money. It refers to the amount charged by a lender to a borrower for the use of assets; it can also be the amount depositors receive or expects to receive on account of depositing their money in a financial institution for a period of time. Interest rate is usually expressed as a percentage of the principal. The assets that could be borrowed include, cash, land, vehicles, building, work animals etc.

Interest rate can be further understood when it is decomposed into saving rate, lending rate and discount rate. Saving rate is the rate of interest paid by a financial institution on the amount deposited with for a stated period of time in accordance with the agreed terms. Lending rate is the interest rate charged by a lender. It is the interest rate which Deposit Money Banks charge their customers on loans extended to them. This rate charged by a lender must cover the cost of funds, a risk premium, loan administration and serving costs including a profit for the owners of the lending entity (Lee, Boehlje, Nelson, & Murray, 1980).

On account of the financial intermediation role banks play in resource mobilization and allocation, deposit money banks are an important channel for the transmission of the Central Bank of Nigeria's interest rate polices. To be in business, banks and other financial intuitions pay interest on deposits and also charge interest on loans and advances to borrowers. The positive difference between the deposit interest rates and the lending interest rates keeps the banks and other financial institutions in business. Discount rate is the interest rate charged when a Deposit Money Bank borrows money from the Central Bank.

Computation of Interest Rate

According to Horne (1980), unlike interest rates on such money-market instruments as treasury bills, most business loans are determined through personal negotiation between the borrower and the lender or the lenders as the case may be. It is therefore necessary to equip farmers with the knowledge that interest rate on loans is subject to negotiation. Furthermore, the amount paid as interest on a loan depends on the method adopted in calculating the interest rate.

There are two ways in which interest on loan may be calculated (Horne, 1980). They are on collect basis and on discount basis. When the computation is on collect basis, the interest is paid at the maturity of the note. For example, if we have a loan of N10,000 at 7% interest rate for a year, the effective rate of interest on a collect note basis is: N700/N10000 multiplied by 100. This gives us 7.00%. When the calculation is on a discount basis, the effective rate of interest is obtained as follows: N700/N9300 multiplied by 100. This gives us 7.53%.

It is obvious that the effective rate of interest is higher on discount basis than on collect basis. So when the Deposit Money Banks require the borrower to maintain demand-deposit balances before loan can be granted or when lenders approve a certain amount of loan and on disbursement pay an amount less than that approved but base the calculation of interest on the gross amount approved, the implication should be clear to the discerning borrower.

According to Horne (1980), minimum demand-deposit balances Deposit Money Banks require a borrower to maintain at a bank in direct proportion to either the amount of funds borrowed, or the amount of commitment are known as compensating balances. If the compensating balance does not require the borrower to maintain balances above those that it ordinarily maintains, such a requirement does not raise the effective cost of borrowing. On the other hand, if balances over and above the ordinary have to be maintained, then the effective cost of borrowing is increased. For example, if we borrow N1million at 7% and we are required to have N100,000 more in balances than we would ordinarily, then we would have to use only N900,000 of the N1million loan facility. As a result, the effective annual interest cost is N70,000/N900,000=7.78% and not 7%.

In a similar fashion, if a borrower is got approval for a loan facility of say N100,000 at 10% per annum and is required to part with N1000 upfront as administrative charges and eventually gets N99,000. If the borrower is required to pay back the principal value of N100,000 at the expiration of the loan term, then the effective annual interest rate is 10.10% and not 10%.

In any case it should be reemphasized that interest rates are negotiable and when agreed upon becomes binding on the relevant parties. The discerning borrower should aware what he or she is paying for. The actual amount paid is in turn dependent on the method adopted in the determination of the interest. Let us look simple interest rate calculation.

Simple Interest Rate

Simple interest is computed by multiplying the interest rate by the principal by the duration of the loan (number of periods). In this calculation, interest on interest is not included as the interest based on the original principal. Mathematically, it is represented as shown in Figure 2.

Figure 2. Simple interest rate formula

$$\text{Simple Interest} = P \times I \times N$$

Where: P, is the loan amount; I, is the interest rate expressed in percentage; and N, is the duration usually in years.

Simple interest calculation is generally applicable to determine the interest charge for short-term loans and only one payment of interest is made at the maturity of the loan. When the simple interest is determined for a period less than a year and in months, the time in years is placed as a fraction with the numerator being the duration of the loan in months and the denominator the months in a year. This same procedure is also applied for determining interest rate for periods less than a month and in days.

For example, the interest rate on N1000 for a year at 8% interest rate is equal to N80. If the duration is for 6 months, the computation is thus:

Simple Interest = 1000 X 0.08 X 6/12 = N40

If the duration is a day, the computation is thus:

1000 X 0.08 X 1/365= N0.22 (approx)

Compound Interest

Compound interest can be thought of as "interest on interest". It is interest calculated on the original principal and also on the accumulated interest of past periods of a deposit or loan. Compound interest is involved whenever the interest is converted at specified time intervals to principal. The time interval can be annually, semiannually, quarterly, monthly or daily as the case may be. It should therefore be understood that compound interest occurs when interest is paid or accrued more than once in the duration of a particular loan term.

On account of the time value of money, money received today has greater value than that received in the future. In order to account for time value of money in investment analysis, all relevant cost and returns that occur in the future are converted to their present values for valid and reliable comparisons. The term present value refers to the value today of a payment to be made at a specified time in the future. The term future value refers to the amount to be paid in a specified future time period.

Compound interest rates help us understand the relationship between present value and future value. This is easily seen in the example that follows. Assuming that N100 deposited in a bank yields 10% annual compound interest rate. At the end of Year1, the amount will grow to N110. At the end of Year3, the amount will grow to N133.1. The mathematics is as shown below:

Initial deposit N100
Yield on N100 in Year1 10
Subtotal at end of Year1 110
Yield on N110 in Year2 11
Subtotal at end of Year 2 N121
Yield on N121 inYear3 12.1
Grand Total at end of Year 3 133.1

Mathematically the amount at the end of the third year is:

$$100(1+0.10)(1+0.10)(1+0.01) + N131.1$$

Alternatively we can state:

$$100(1+0.10)^3$$

The foregoing presentation can be generalized as

$$S=P(1+i)^n$$

Where S= future amount at the end of n period
P= amount invested at period n
i=interest rate
n=number of time period usually in years

This is the general equation for compounding. By way of formal definition, compounding can be construed as the process of finding the future value of a given present value. It is clearly the opposite of discounting (The process of determining the present value of a given future value.). Compounding is used to determine the future value of some amount in the present especially in investment analysis. The spread between the cost of funds and the interest rate charged to borrowers represents one of the main sources of profit for most financial institutions.

The effect of compounding on the amount of interest paid actually depends on the frequency of the conversion of the interest amount to the principal

Table 2. Effect of frequency of compounding on value of interest paid on a n1000, 7% loan facility repaid at end of 1year

Item 3	Loan1	Loan 2	Loan
Frequency of Compounding	Annually	Semiannually	Quarterly
Principal Amount	N1000	N1000	N1000
Accumulated Interest at end of:			
3 months	0	0	17.50
6 months	0	35	17.81
9 months	0	0	18.12
12 months	70.00	36.23	18.44
Total interest paid	70.00	71.23	71.86
Total due at end of 1year	1071.23	1071.86	1070.00
Rate per conversion period (Compound rate)	7%	3.5%	1.75%
Effective annual rate	7.00%	7.12%	7.18%
Nominal rate	7%	7%	7%

Source: Adapted from Lee, Boehlje, Nelson, & Murray, 1980

amount. By way of illustration, let s compare three one-year N1000 loans with a nominal interest rate of 7%. Table 2 depicts the effect of frequency compounding on value of interest paid on a N1000, 7% loan facility repaid in a single payment at end of one year.

Table 2 shows that the amount of interest charged increased with the increase in the frequency of compounding. Lee, Boehlje, Nelson, & Murray (1980), employed the table to clearly depict this effect. For the purpose of showing working, let us employ the relevant formulas to drive home the point home. Using the expression: $S=P(1+i)^n$

Where S= future amount at the end of n period

P= amount invested at period n
i=interest rate
n=number of time period usually in years

For the computation on an annual basis, we have:

S_1 =N1000(1+0.07)= N1070.

If the time is n years, the future amount at the end of n years will be

S_n = N1000$(1+ .07)^n$ or more general generally: $S_n = X_o(1+r)^n$. All the terms are as previously defined.

The positive relationship between the terminal values of compound interest and the interest rate for different periods of compounding becomes more obvious when the computation is on a less than annual basis. For instance, on a semi-annual basis, the amount at the end of the period is:

$S_{½}$ = N1000 (1+ .07/2) = N 1035. At the end of the year it is:

S1=N1000 $(1+.07/2)^2$ = N 1071.225.

Note that if the interest was paid once a year, the value is N1070. If the interest was paid twice a year, the value is N1071.225. The difference N1.225 is attributable to the fact that during six months, interest is earned on N1.225 interest paid at the end of the first six months. The more times during a year that interest is calculated and accrued on a loan, the greater the terminal value at the end of a given year.

According to Horne (1980), the general formula for solving the terminal value at the end of year n where interest is paid m times a year is:

$$TV_n = X_o (1+r/m)^{mn}$$

For example, in Table 2 the quarterly computation of the compound interest at the end of one year gives:

S_1=N1000$(1+0.07/4)^4$=1.07185903*1000=N1071.86. This value is higher than those of semiannual and annual compounding.

INTEREST RATE AND AGRICULTURAL PRODUCTIVITY IN NIGERIA

The capacity to be a major player in the global economy is dependent on the growth and productivity of the different sectors of the economy including agriculture. Interestingly, growth and productivity can either be enhanced or impaired by the monetary policies of the government.

According to Lee, Boehlje, Nelson, & Murray (1980), monetary policy is concerned with the volume of money and credit in the economy and the conditions under which they are available. When the Central Bank manages the money supply to influence the economy, economists call it monetary policy (Szulczyk, 2014). Drummond and Godwin (2004) view monetary policy as deliberate changes in the money supply in an effort to affect the macro economy. By deliberately regulating the supply of funds and terms of availability, the Central Bank can significantly influence the level growth and productivity in the economy in general and the agricultural sector in particular.

One of the major obstacles in increasing productivity in the agricultural sector is the absence of viable credit facilities. Where credit facilities are available, the average farmer finds it difficult in accessing them. Difficulty of access is further compounded by high interest rates when a loan application manages to scale through.

The problem of high interest rate is weightier on the smallholder farmers than on large scale and established farmers. For instance, the lending interest rate for Deposit Money Banks ranges from 22-34%. If the rate is referred to as being high, it is spectacularly prohibitive in the informal sector. The informal sector rates are somewhat beyond the scope of this book. It is pertinent to point out how the lending rates are fixed.

The Monetary Policy Rate (MPR) is fixed by the Central Bank of Nigeria. Also referred to as benchmark rate or mother of all rates, the MPR is the rate at which the Central Bank of Nigeria, in performing its duties as the lender of the last resort, lends to Deposit Money Banks. According to Hummel (n.d), a central bank determines the lowest possible interest rate in an economy, called the prime rate and charges this rate on loans to Money Deposit Banks. Money Deposit Banks in turn charge each other a similar rate on loans. Consequently, Deposit Money Banks charge their customers a higher interest rate, (higher than the prime rate). However, this interest rate charged goes up and down with the prime rate.

The objective of price stability of the Central Bank of Nigeria influences the level at which the MPR is fixed. The CBN sets the lower and the upper levels to guide all market interest rates. The CBN controls the amount of money in circulation at any point in time through the manipulation of the MPR. Other things being equal, when the MPR is raised by the Central Bank, the money supply contracts. Conversely, when the Central Bank lowers the MPR, the money supply expands. According to Duffy (nd[1]), "Tight" money slows economic activity down. The slowing down in economic activity is on account of high interest rate. On the other hand, "loose" money speeds up economic activity. The MPR is used to control the amount of money in circulation, grow the economy and also check inflation. The MPR has been set at 14%

Fundamental to a growing economy, low interest rates encourage borrowing and investing. On the other hand, high interest rates bring about prudence and tend to limit investment. In a formal economy, the quantity of loanable funds demanded by consumers is inversely related to the interest rate. With the MPR and the lending rates of Deposit Money Banks set at 14 and 34% respectively, the prevailing interest rates are clearly prohibitively high. There is therefore the need to reduce the interest rates and thereby release more loanable funds to the economy including the agriculture sector of the economy.

According to Drummond & Godwin (2004), monetary policy is certainly important to the agricultural sector. A tight monetary policy always hurts the farmer because high interest rates will increase cost while unemployment and recession will reduce the demand for farm products. On the other hand, lower interest rates tend to stimulate the economy including the agricultural sector and so give rise to growth and enterprise expansion.

In spite of the calls for the Central Bank of Nigeria to reduce the interest rates in the face of the current economic downturn, the Bank is adopting a different approach. The Central Bank of Nigeria has for 12 months held rates steady for the purpose of controlling inflation. It should be stated that though the inflation rate is on a gradual decline, it is still high at the current level of 16.25%. The Central Bank of Nigeria argues that if interest rates are lowered, more money will eventually be pumped into the economy and aggravate inflation as a consequence. According to Duffy (nd[2]), inflation flares up whenever too much money chases too few goods. To curb inflation, the Central Bank pulls the string by reducing the amount of money in circulation.

The Central Bank has to walk a very tight rope in this regard. If the rates are left at the present high rates, demand for loanable is constricted. If the rates are lowered, inflation tends to be aggravated as a consequence. How well the tight rope is walked for it to stimulate the economy and so give rise to agricultural growth and enterprise expansion should be a measure of the effectiveness of the Central Bank of Nigeria in managing the agricultural economy in particular.

Walking the tight rope becomes a bit easier when it is realized that the nature of inflation currently witnessed in Nigeria is not that of too much money chasing too few goods. The type of inflation currently prevailing in the Nigerian economy is brought about by high cost of production. High cost of production finds expression in transportation, energy and infrastructure.

The Deposit Money Banks are confronted with realities of high cost of production in terms of energy, fixed interest rates and the cost of risk of doing business. The Deposit Money Banks in turn pass on the costs to their customers. Some customers in some sectors of the economy can more easily pass on the costs to their clientele than others.

Small-holder farmers in particular and the agricultural sector in general are not particularly equipped to pass on the cost of doing business to their customers. This fact reinforces the argument lowered interest rates and also targeted financial assistance with the aim of improving agricultural productivity.

SOLUTIONS AND RECOMMENDATIONS

It is recommended that farmers should be aware of the costs associated with the funds they employ in their farm businesses. Furthermore, whether an enterprise is financed using debt and or equity; cost is associated with the use of the funds. In employing funds for running an enterprise, the farmer should strive to repay the borrowed funds and make a margin of profit.

FUTURE RESEARCH DIRECTIONS

Some farmers do not borrow erroneously thinking that it a way out of financial fragility. Those who are courageous enough to borrow out of indiscretion incur so much cost on the borrowed funds that they sink into further debts and therefore see borrowing as being unprofitable. There is therefore the need to conduct research into why farmers borrow and why they do not. Furthermore, there is need for an empirical investigation into the profitability of borrowing especially among small-scale farmers.

CONCLUSION

Cost of capital is the composed of the costs of the various components of financing an enterprise. Cost of funds is the interest rate paid by financial institutions for the funds that they deploy in their business. For borrowers, cost of funds is determined by the costs they must incur to gain access to funds. It is therefore necessary to equip farmers with the knowledge that interest rate on loans is subject to negotiation and that the amount paid as interest on a loan depends on the method adopted in calculating the interest rate. Cost of funds is borne by the farmer borrower with the expectation of making positive returns on the invested funds. Farmers in particular and investors in general should constantly keep track of their costs and returns of capital to see on which side the scale comes down.

REFERENCES

Anon. (2014). *Accounting and Finance for Business Analysis*. DELTACPE LLC.

Duffy, F. (n.d.a). *What Is the Monetary Policy Rate?* Retrieved July 28, 2017, From http://www.ehow.com/about_5149730_monetary-policy-rate.html

Duffy, F. (n.d.b). *Examples of Tight Monetary Policy*. Retrieved July28, 2017, From http://www.ehow.com/about_5296165_examples-tight-monetary-policy.html

Drummond, H. E., & Godwin, J. W. (2004). *Agricultural Economics* (2nd ed.). Prentice Hal Inc.

Fernandes, D. (2014). *Finance for Executives: A practical guide for managers*. McGraw Hill.

Hummel, C. (n.d.). *Monetary Vs. Fiscal Policy*. Retrieved July28, 2017, From http://www.ehow.com/facts_5241421_examples-monetary-policy.html

Horne, J. V. (1980). *Financial Management and Policy*. Prentice/Hall International London.

Lee, W. F., Boehlje, M. D., Nelson, A. G., & Murray, W. G. (1980). *Agricultural Finance* (7th ed.). The Iowa State University Press.

Maharshi Dayanand University. (2004). *Financial Management*. Maharshi Dayanand University.

Szulczyk, K. R. (2014). *Money, Banking, and International Finance*. Available online: www.money-banking-and-international-finance.pdf

Staff, I. (2017, March 6). *Cost of Debt*. Retrieved April 16, 2017, from http://www.investopedia.com/terms/c/costofdebt.asp#ixzz3qDe2EyYv

Staff, I. (2015a, December 30). *Cost Of Equity*. Retrieved November 16, 2016, from http://www.investopedia.com/terms/c/costofequity.asp#ixzz3qDXJNFDK

Staff, I. (2015b, September 29). *Weighted Average Cost Of Capital - WACC*. Retrieved November 16, 2016, from https://www.investopedia.com/terms/w/wacc.asp#ixzz3tdarmPdM

Staff, I. (2015c, October 26). *Cost Of Funds*. Retrieved November 16, 2016, from http://www.investopedia.com/terms/c/costoffunds.asp#ixzz3qDRrGJ63

Chapter 11
Valuation and Depreciation of Farm Assets

ABSTRACT

Assets, especially fixed assets, are obtained at a cost and are used in the course of more than one production period. In line with the method of acquisition of the relevant asset, the need arises that their accurate values are appropriately spread over the lifetime of the asset in question and over the relevant production period. It is against this background that this chapter is devoted to highlighting the concept of asset valuation. The discussions are based on a review of relevant literature. It is concluded that the realistic valuation of assets involved in a farm enterprise is essential in ensuring that enterprise is a going concern. It is recommended that relevant contributions of the farm assets including hoes, machetes, head pans, wheelbarrows, and all should be meticulously and rigorously factored into the costs and returns of the production cycles to give a complete idea of the enterprise as a going concern. Furthermore, farm asset such as land that appreciates in value should be realistically incorporated in the valuation of the production in the farm venture.

INTRODUCTION

Entrepreneurs in general and small-scale farmers in particular make use of assets in their production activities. The assets are obtained at a cost and some, especially the fixed assets, are used up in the course of more than one production period. In some cases, it is popular to think that small-holder

DOI: 10.4018/978-1-5225-3059-6.ch011

farmers use little or no fixed assets and as a result analysis should be based on the gross margin. The need is therefore evident that proper and realistic values are assigned to the assets. Realistically determined values of the assets will obviously guide the discerning business minded acquiring farmer or entrepreneur in reaching a decision on whether to purchase or to the asset in question and at what cost in relation to the expected returns to the overall profitability of the enterprise. In line with the method of acquisition of the relevant asset, the need also arises that their accurate values are spread by degrees over the lifetime of the asset in question and over the relevant production period. This chapter is therefore devoted highlighting the concept of asset valuation. It also delves into the methods of asset valuation with particular emphasis on depreciation.

THE CONCEPT OF VALUE AND ASSET VALUATION

The basis of economics is exchange of goods and services for money. A seller sells say cassava tubers to a consumer and buys vitamin C enriched cassava stems from the office of the Agricultural Development Programme. Both the farmer seller of cassava tubers and the extension agent seller of cassava stems have acquired something they wanted more than what they had.

According to Drummund and Goodwin (2004), a buyer places a greater value on the article bought than on the money exchanged for it. The same is also true of a seller: the seller places greater value on the money received than on the article sold. If it were not so, neither the buyer nor the seller would have allowed the transaction to take place.

According to Hingley and Osborn (n.d), quite frequently the term value is automatically equated with cost or monetary equivalent. They went further to argue against automatically equating value with a monetary equivalent citing the technique of value analysis which differentiates at least four types of value: use value, exchange value, esteem value and cost value.

Acceptably, value is different from cost and cost is different from price. Economics, especially in its policy applications essentially deal with value system of individuals and groups of individuals (Drummund & Goodwin, 2004). In a philosophical sense, value is what we think should be the case. This is against what we know to be the case (facts), and what we think to be the case (beliefs). As an action word, value means to place worth on a specific item, entity, thing or event. As a noun, it is a fair equivalent, the worth of something in terms of the amount of other things for which it can be exchanged or in terms of a medium of exchange as opposed to "automatic equivalent" of something. In economics, value can be construed as equivalent worth or return in money, material worth, and services. In the views of Drummund and Goodwin (2004), price refers to a per unit concept; for example, the price per liter of petrol. Cost refers to a total concept of price multiplied by quantity the quantity purchased. For example, the cost to fill up the tank of a petrol generator set.

In all it makes huge sense relating value to monetary equivalents to arrive at a relative sense of worth. According to Lipsey (1989), we value our wealth, our incomes, what we buy and what we sell in monetary terms. So, money prices measure economic activities. Expressing values in monetary equivalents enables anyone to compare two or more things and thereby reach a decision. Because total agreement about values is never anticipated; so there must be some procedure for determining which values should dominate. The discussion under the valuation of assets is therefore based on relating the worth of an asset to the monetary equivalent at a specific point in time.

Valuation is the term used to refer to the process of determining the monetary value of a venture (Duening & Sheril, 2006). It is the determination of the value of resource in a certain type of employment. Asset valuation involves determining a realistic measure of the current value of assets in an enterprise (Olukosi & Erhabor, 1988). Asset valuation involves assessing and thereby assigning monetary value to a property or claim owned by an individual or an enterprise. The asset can be current asset or fixed asset. According to Gittinger (1982), current asset consists of cash and items expected to be converted into cash within a reasonably short period usually a year. An example of current asset is cash at hand, seed and fertilizers etc. On the other hand, fixed asset refers to durable items of comparatively long-life span that are used by an individual or an enterprise for the production of goods and services. More often than not, valuation of assets focuses specifically on the fixed assets.

For one reason, fixed assets are employed in an enterprise at purchase price and usually reduced overtime by some deprecation provision made on the initial (Hingley & Osborn, n.d). For another, current assets are usually reflected in the accounts at current values subject to any specific adjustment such as bad debts

According to Hingley and Osborn (n.d), business and share valuations are carried out primarily for one or more of the following purposes. For unquoted businesses, valuations are carried out for tax purposes and for determining a selling price for the business as a going concern. For quoted businesses, valuation is done to confirm the need or otherwise of its own market valuation; to determine the price at which new shares or rights issues may be made; and by one firm of the shares of another in readiness for purchasing as an investment or take over proposition.

It should be added that in both unquoted business and quoted businesses, valuations can also be carried out to determine to relative contribution of a capital asset to the production processes over the life span of an asset especially fixed assets. Discussions here are as they relate to unquoted businesses since the small-scale farmers are in focus. Furthermore, the valuations are as they relate to determining the relative contributions of a fixed capital asset to the production processes over the life span of the relevant assets.

VALUATION TECHNIQUES

The basic approach to valuation is by reference to either what the business owns, that is assets or to what the business can create, that is earnings. In this chapter we are interest in assets valuation with a view to determining how the value of the assets are, by degrees, translated to the production processes over the entire life of the asset. According to Olukosi and Erhabor (1988), valuation techniques include valuation at cost, valuation at market price, valuation at net selling price, valuation by reproductive value and capitalization of earnings. Let us look at the different techniques and connect the discussion to how the typical small-scale farmer can keep track of the relative contributions of the farm assets to production activities in the enterprise.

Valuation at Cost

This involves valuing the farm inventory based on the actual amount invested on the asset when it was initially acquired. This approach has obvious flaws. The values of assets are in constant state of flux. While some appreciate, for instance land, some depreciate, for example buildings and machinery. The obvious relevance of this approach is that the amount invested on the asset when it was originally bought provides a basis for the determining its value at any point in time.

Valuation at Market Price

This involves assigning value to the asset in question on the basis of the current market price of the asset in question. This has its drawbacks as it results in an over-estimation or under-estimation of the value of the asset depending on the state of the economy. This approach of valuation at market price is usually employed when valuing land by finding out what similar lands in the vicinity are going for. It has been established that what the buyers are willing to pay and the sellers willing to accept is known as the fair market value for the item in question.

Valuation at Net Selling Price

Valuation at net selling price is arrived at when the cost of selling an item (advertisement, transportation, agency fees etc) is subtracted from market price of the item.

Valuation by Productive Value

Valuation by productive value entails the determination of what it would cost to produce a good or service at present price and under present methods of production. According to Olukosi and Erhabor (1988), this method is useful for long term assets and has little or no application for current assets.

Valuation by Contribution to Production

Under the valuation by contribution to production, the focus is on the value of increase in production as a result of a unit increase in a relevant input. In this approach, marginal analysis in the value added is employed.

Valuation by Capitalization of Earnings

Valuation by capitalization establishes the value of the enterprise by dividing the expected business economic returns by the capitalization rate (ValuAdder, n.d.). The capitalization rate is the rate of return on a real estate investment property based on the income that the property is expected to generate (Staff, 2016). The capitalization rate of an investment can be determined by dividing the investment's net operating income by the current market value of the property. The net operating income is the annual return on the property minus all operating costs. The formula for calculating the capitalization rate can be represented as:

Capitalization Rate = Net Operating Income / Current Market Value.

The capitalization rate is usually expressed as a percentage and is also often known for short as the cap rate. For example, if a piece of land is bought for N800,000 and the land is expected to generate N100,000 per year after operating costs, the capitalization rate for his investment is 12.25% (N100,000 / N800,000 = 0.125 = 12.50%). This means that on a yearly basis this property yields 12.5% of the value as profit. This illustration assumes that all factors in the calculations will remain constant. In reality, this is obviously not the case. The cap rate represents the risk associated with the receiving the benefit in the future. According to Olukosi and Erhabor (1988), the valuation by capitalization of earnings approach is more suitable for durable capital such as land, whose contribution to the production process extends to many years into the future.

In all valuation techniques, it should be borne in mind that the farmer or the analyst as the case may be is expected to come up with a fair and realistic worth of the asset or the venture in question. According to Ejiogu (2002), certain principles are involved in arriving at the values of items included in the balance sheet. These principles, also known as valuation rules are in general accord with the fundamental accounting concepts and are as follows:

1. Assets and liabilities are valued on the assumption that the enterprise is a going concern;
2. Accrual concepts are used, that is revenues and costs are accrued — recognized as they are earned or incurred — and not as money is received or paid; and
3. Conservatism which avoids placing too high a value on any item.

Some assets lose value with time. For example farm tools and machinery lose value over time whether in use or not. On the other hand, some assets gain value over time. An example of an asset that appreciates in value over time is land. In the next section we look at concept of loss in value of capital assets over their life time.

DEPRECIATION

When it is purchased, a fixed asset is initially valued at cost. Each year thereafter it is depreciated reflecting the fact that they wear out and must eventually be replaced. According to Olukosi and Erhabor (1988), depreciation is the loss in value of a capital asset over time on account of age, obsolescence and wear and tear. It can also be referred to as the continuous and gradual loss of the book value of fixed assets. It can also be referred to as the continuous and gradual loss of the book value of fixed assets. It can further be described as the payment in installments by periodically setting aside some sum of money, arrived at based on purchase price of the relevant fixed asset, its useful life span and its salvage value with a view to generating a replacement value of the asset at the end of the asset's life span. In the course of time, most fixed assets (excepting land) are reported in the books at progressively lower values (i.e., book value). Eventually, the assets in question are regarded to have no more value or to have reached a value, below which it cannot be economically reduced any further (salvage value). Depreciation is a function of useful life of the asset (time), its purchase price and its salvage value. So instead of having to commit a large chunk of money to replace a worn-out asset, on account use and time, the annual series of cost set aside as depreciation is used. Depreciation not only spreads the original cost of an asset in question over its useful life span, it also gives an idea of how much the asset contributes to production over the years.

Methods of Calculating Depreciation

There are three popular methods of calculating depreciation. They are the straight line method, the reducing balance method and the sum of the years digits.

Straight Line Method of Depreciation

The straight line method of calculating depreciation involves making provision for a fixed payment of a proportion of purchase price of the fixed item over each accounting period of the asset's useful life span. The fixed amount also known as the provision for depreciation is determined using the formula:

$D = P - S/N$

Where D is the depreciation amount per year; P is the purchase price of the asset in question; S is the salvage value and N is the number of useful life span of the asset. This method of determining depreciation is quite simple. Its simplicity tends to account for its general use.

Reducing Balance Method

The reducing balance method, also referred to as the declining balance method is an approach in depreciation computation which is based on a fixed percentage of the book value of the asset as reduced by the previous year's provision. The salvage value of the asset does not come into the calculation as it is not subtracted from the purchase price before calculation. The fixed rate of depreciation is multiplied by purchase price of the item. The value obtained from the multiplication, call it a_1, is the depreciation value for the year in question. Subtract *b* from the purchase price; call the value obtained *b; b* is the remaining balance for further calculation of the depreciation for the subsequent year. In the subsequent year, multiply *b* with the fixed rate of depreciation and let the value be $a_{2.}$ Then a_2 is the depreciation value for the year in question. Subtract a_2 from b and let the resulting value be *c; c* is now the remaining balance. The process is continued until the salvage value of the asset is obtained.

For instance, if an asset has purchase price of N5000 and a fixed rate of depreciation of 20% the depreciation value is obtained by multiplying the purchase price with the fixed rate. The calculations are as shown in Table 1.

Table 1. Reducing balance method

Year	Annual Depreciation	Remaining Value
1	20% of 5000 = 1000	4000
2	20% of 4000 = 800	3200
3	20% of 3200 = 640	2560
4	20% of 2560 = 512	2048
5	20% of 2048 = 409.6	1638.42
6	20% of 1638.4= 327.68	1310.74
7	20% of 1310.74 = 262.15	1048.59
8	20% of 1048.59 = 209.72	838.87
9	20% of 838.87 = 167.77	671.1
10	20% of 671.1 = 134.22	536.88

As shown in Table 1, the annual depreciation value is higher in the early life span of asset than in closing years. The choice of the fixed rate of depreciation is at the predilection of the farmer or the analyst and is usually based on depreciation rate of similar enterprise in the locality.

Sum of Years Digits Method

Using the sum-of-years digit method, annual depreciation is determined by multiplying a fraction by the purchase price of the asset being depreciated (Olukosi & Erhabor, 1988). The fraction is obtained by summing up the digit 1 to *n* for an asset with n useful years. For instance, if the number of years of the life span of an asset is say 5, the sum of the digits is 1+2+3+4+5= 15. The figure 15 forms the denominator for the fraction while the numerator is the remaining use life span of the asset as at the period of calculation. In this case, for the first year the fraction 5/15, the second is 4/15 and so on. For an asset with purchase price of N5000 and a salvage value of N500 and a useful life span of 5 years, the depreciation schedule using the sum-of-years digits is as shown in Table 2.

According to Olukosi and Erhabor (1988), the calculation can be done using the formula:

$$D_a = 2W\ (n\text{-}a + 1)/n(n + 1)$$

Where D_a = depreciation in year a
W= value be depreciated
n = number of years for depreciated value
a = year under consideration

Using the formula, the first year is calculated thus:
Substituting the values we have; for the first year

Table 2. Sum of years digits method

Year	Annual Depreciation	Remaining Value
1	5/15 x (5000- 500) = 1500	3500
2	4/15 x (5000- 500) = 1200	2300
3	3/15 x (5000- 500) = 900	400
4	2/15 x (5000- 500) = 600	300
5	1/15 x (5000- 500) = 300	300

$D_1 = 2(5000 - 500)(5-1+1)/ 5(5+1) = 45000/30= 1500$

For the second year

$D_2 = 2(5000 - 500)(5-2+1)/ 5(5+1) = 36000/30 = 1200$ etc

We have thus far looked the different methods of determining the depreciated value of an asset that depreciates in value over time and use. On the other hand, some assets gain value over time. An example of an asset that appreciates in value over time is land. The question therefore arises as to how to realistically and fairly incorporate the value of land over production cycle. The approach is simple. According to Ejiogu (2002), the value of land for accounting purpose is usually based on the price of similar piece of land in the specific locality. Where necessary, adjustments are made to keep the value in line with current estimates. With particular reference to the particular value of land that can be attributed to have contributed a current production cycle, that can be determined in line with opportunity cost of renting a similar piece of land for the production period.

SOLUTIONS AND RECOMMENDATIONS

It cannot be overemphasized that a farmer should keep up to date, relevant and reliable records. Records are essential in enabling the farmer know the value of the enterprise at any point in time. So valuation is as realistic as the records on which it is based. Reliable farm records are the first line of solution to the problem of valuation. The farmer analyst should also be guided by the fundamental valuation principles of not placing too much or too little value on any item. Bringing business sense to small scale agriculture partly entails valuing relevant resources to know if the enterprise is growing or not. The relevant contributions of the farm assets including hoes, machetes, head pans wheelbarrows and all should be meticulously and rigorously factored into the costs and returns of the production cycles to give a complete idea of the enterprise as a going concern. Farm asset such as land that rather appreciates in value should be realistically incorporated in the valuation of the production in the farm venture.

FUTURE RESEARCH DIRECTIONS

It is popular misconception to view farmers especially small-holder farmers as operating with little or no fixed asset; and so small-holder farm analysis should be based on the gross margin. This thinking is fundamentally flawed as the small-holder farmers are known to own and employ such simple farm as hoes, machetes, head pans wheelbarrows and all; they obviously own some durable farm structures. Under collective or individual ownership arrangements, some own heavy farm machine such as tractors. It therefore suggested that the relative contributions of the simple farm tools should be investigated as opposed to ignoring as being inconsequential in profitability analysis of the farm enterprise.

CONCLUSION

Enterprise expansion can come about through increase in value of the resources of the enterprise. Realistic valuation of all the assets involved in the farm enterprise is essential in enduring that enterprise is a going concern.

REFERENCES

Drummond, H. E., & Godwin, J. W. (2004). *Agricultural economics*. Prentice Hall.

Duening, T. N., & Sheril, W. W. (2006). Entrepreneurship exploring entrepreneurship from a business process perspective. Atomic Dog Publishing.

Ejiogu, A. O. (2002). *Financial analysis of small-holder rice production enterprise in Imo State* (Unpublished doctoral dissertation). University of Nigeria Nsukka.

Gittinger, J. P. (1982). *Economic analysis of agricultural projects*. The Economic Development Institute of the World Bank, The John Hopkins University Press Baltimore.

Hingley, W., & Osborn, I. (n.d.). *Financial management*. Institute of Financial Management.

Lipsey, R. G. (1989). *An introduction to positive economics*. London: English Language Book Society.

Olukosi, J. O., & Erhabor, P. O. (1988). *Introduction to farm management economics principles and applications*. Zaria, Nigeria: Agtab Publishers Limited.

Staff, I. (2016, October 6). *Capitalization Rate*. Retrieved November 16, 2016, from https://www.investopedia.com/terms/c/capitalizationrate.asp?ad=dirN&qo=investopediaSiteSearch&qsrc=0&o=40186

ValuAdder. (n.d.). *Capitalization Rate*. Retrieved November 16, 2016, from http://www.valuadder.com/glossary/capitalization-rate.html

Chapter 12
Agricultural Risk Management and Insurance

ABSTRACT

Risk management in agriculture can be described as the art and science of choosing among alternatives aimed at reducing the adverse effects of foreseen and unforeseen variability in the production process. Because agriculture is a biological activity, the practice is prone to risks and uncertainty. The need is urgent and cogent to continue to draw attention to the risky nature of the practice of agriculture and thereby proffer effective risk management strategies. This chapter therefore focuses on the basic concepts of agricultural risk management and insurance as they relate to agricultural finance and enterprise expansion. The discussions are based on a review of relevant literature. The chapter concludes that granted that the practice of agriculture is a private sector activity, the need for institutional support, especially in the area formal agricultural insurance schemes, cannot be overemphasized. It is recommended that relevant institutions involved in agricultural insurance as a formal risk-management strategy should ensure that famers obtain due compensation for their insured enterprises in the event of the relevant catastrophe in the farm enterprise.

DOI: 10.4018/978-1-5225-3059-6.ch012

INTRODUCTION

This chapter focuses on the basic concepts of agricultural risk management and insurance as they relate to agricultural finance and enterprise expansion. It defines the relevant concepts, connects them to decision making in risk management against the background of agriculture being a risky business. The chapter is summarized with conclusion and recommendations.

BACKGROUND

Risk management in agriculture can be described as the art and science of choosing among alternatives aimed at reducing the adverse effects of foreseen and unforeseen variability in the production process. The foreseen and unforeseen variability in agriculture production process come by way of risks and uncertainty. The ability to mitigate, control or reduce the risks that militate against production and enterprise expansion can be referred to as risk management. The decision to take up agricultural insurance policy comes under the purview of risk management.

Agricultural insurance can be construed as financial instrument employed to transfer agricultural production risks to a third party through payment of periodic premium to an insurance agency by the entity taking up the insurance. According to Knong (2007), the insurance agency is able to pool the risks by accepting appropriate premiums from a large number of clients. The overriding objective of agricultural insurance is to shield farmers against risks through an arrangement that guarantees immediate and due indemnity required to keep the farmers in business in the event of any mishap.

According to NAIC (2017), agricultural insurance as a concept refers to transfer of the risk of the agricultural enterprise from the farmer to others for a price which is stated as an insurance premium. Agricultural insurance essentially shares the risks attendant to the production process. Agricultural insurance is one of the risk management services available to farmers.

Interest in agricultural insurance can be justified on the basis of the myriad of risks and uncertainties faced by investments in agriculture. Agricultural activities are by nature prone to risks and uncertainty. Against the background of the emphasis of the diversification of the Nigerian economy away from oil, there is the need to consciously mitigate, control and reduce the inherent risks associated with agricultural investments. This is where the principles and practice of agricultural insurance come in.

RISKS AND UNCERTAINTY IN RELATION TO AGRICULTURAL INSURANCE

One of the typical characteristics of agricultural enterprise is that it is beset with considerable risks and uncertainties. These attendant risks and uncertainties are on account of agricultural enterprises being predominantly dependent on nature. Risk and uncertainty define the probability of exposure to occurrences that are capable of engendering losses. While risk refers to the occurrence of adverse events of known probabilities, uncertainty refers the occurrence of adverse events of unknown probabilities. Risk and uncertainty both define a situation a varying number of outcomes are possible.

According to Cervantes-Godoy et al., (2013), risk refers to a combination of the probability of an event and its consequences. According to Teweldemedhin and Kapimbi, (2012), and Nto et al., (2014), risk refers to the probability of occurrence of an event or condition that, if it occurs, would have a negative or positive effect on one or more project objectives. According to Surbhi (2016), in the ordinary sense, risk is the outcome of an action taken or not taken, in a particular situation which may result in loss or gain. The probability of winning or losing something is known as risk. The terms risks and uncertainty are usually associated with the exposure to outcomes that can result in losses. So, reference to the probability of positive effects of the occurrence of risk or the probability of winning being a component part of the definition of risk should be interpreted with caution. The interpretation can be found in the maxim that the higher the risk, the higher the returns in the event of gain; the higher the risk, the higher the cost, in the event of loss. It can also be interpreted from the view point of the concepts of downside risk and upside risk. While upside risk refers to the uncertainty of gain, down side risk is associated with loss.

In the financial sense, Surbhi (2016), observed that the meaning of risk is not much different and went on to state that it refers to the probability of actual returns on an investment may not being equal to the expected returns. Risk can be construed as hazard, chance of bad consequences, loss, and exposure to mischance. According to Hingley and Osborn (n.d.), risk arises by virtue of incomplete knowledge of the outcome of a specific course of action. The inability to be certain about the occurrence of an event gives rise to risk. Knight (1921), referred to risk as unknown outcomes whose odds of happening can be measured or at least learned about. For example, in throwing a fair coin, what is not known is whether the coin lands heads or tails. It is known that if we flip the coin, say 100 times, the coin would on average fall 50 times heads and 50 times tails. Thus we know exactly the odds of each of the possible events. According to Guerron-Quintana (2012), this is precisely the essence of risk: We can describe the odds of the unknowns.

Guerron-Quintana (2012), referred to uncertainty as uncertain events that we do not even know how to describe. It is an expression of the degree to which a value or the occurrence of an event is unknown. It is the absence of certainty or something that is unknown. It describes a situation where there is more than one alternative giving rise to a specific outcome whose probability of the outcome is uncertain. The uncertainty derives from insufficient information or knowledge about the condition at hand. Uncertainty cannot be measured in quantitative terms through past models. Therefore, probabilities cannot be applied to the potential outcomes, because the probabilities are unknown. Langemeier (2015), viewed uncertainty as representing a situation in which information on the probability occurrence of the event is not available. When analyzing business plans in terms of the best case, the most likely case, and the worst case scenarios and we do not have probability information available as to how likely each of these scenarios will be in the future, we are dealing with uncertainty rather than risk. Uncertainty is unquantifiable as future events are unpredictable, and information is insufficient.

We have so far attempted to distinguish between risk and uncertainty. The twin concepts of risk and uncertainty have often been used interchangeably in the popular press, but economists have long distinguished between the two. Surbhi (2016), summarized the major differences between the two terms are as follows:

1. Risk refers to the probability of winning or losing something; uncertainty refers to the condition where there is no knowledge about the future events.
2. Risk can be measured and quantified, through theoretical models; on the other hand, on account of the unpredictable nature of future events, it is not possible to measure uncertainty in quantitative terms.
3. The potential outcomes are known in risk; those of uncertainty are unknown.
4. If and when proper measures are taken, risk can be controlled; uncertainty is beyond control.
5. In risk, probabilities are assigned to a set of circumstances which is not possible in case of uncertainty.

TYPES OF RISKS IN AGRICULTURE

Some of the risks affecting agricultural production include the following:

Production Based Risks

Production risks are brought about by the inherently biologically based growth process of crops and livestock. Agricultural production based risks relate to the probability that farm output levels will be lower than expected. The risky nature of the natural growth processes is further compounded by uncertainty caused by variations in climate and the vagaries of the weather. Variations in climate and the vagaries of the weather find expression in drought and flood and in the incidence of pests and diseases. Some of the main sources of agricultural production risks include such vagaries of the weather as drought, flood etc and loss or damage to the crops by pests and diseases.

Price Risks

Price risks also known as market risks refer the risks of market price fluctuations. Market risks may occur when an enterprise's forecasts of market size or price levels are not accurate. Food prices dramatically fluctuate in and out of season, and so make production and planning uncertain. According to Bosede (2010), limited supply of such farm inputs as fertilizer, seeds, chemicals etc give rise to uncertainty in the production and planning process. According to Lee, *et al* (1980), government price support programmes can help reduce price uncertainty. It should be stressed that farm input and farm output price fluctuations are both implicated in price risks. This emphasis is necessary so as to inform and influence policymakers on the need to have a balanced and complete approach to agricultural insurance schemes that guarantee growth and profitability of agricultural production.

Business Risks

Business risk refers to the variability in the net earnings of an enterprise resulting from the inherent nature of the enterprise doing the business. Risks, including business risks are characteristics of agricultural enterprises because of their biological nature and their dependence on clemency of the weather conditions. Furthermore, agricultural commodities are inherently easily perishable. Processing may prolong the shell life of processed products but does not altogether prevent perishability. So, perishability, incidence of pest and diseases and changes in weather and climatic conditions result in agricultural business risks.

Financial Risk

Financial risk is defined by the proportionate relationship between the debt-equity combinations in a production enterprise. According to Horne (1980), the use of debt in an enterprise creates a fixed financial obligation by way of settlement of principal, interest rent and other related commitments. This commitment to suppliers of debt capital employed in an enterprise results in financial risk. As the amount of debt capital relative to equity capital increases, the financial commitment increases and so does the financial risk. Some enterprises are riskier than others. When these projects are undertaken, prospective returns from investments in riskier enterprises are more uncertain than others. An enterprise will be more or less risky depending on the amount of debt in relation to equity in its capital structure.

RISK MANAGEMENT IN AGRICULTURE

The ability to control or reduce risk is referred to as risk management. It involves choosing among alternative actions to mitigate downside risk. Risk bearing ability of the average farmer must essentially be enhanced if agricultural production is to be a going concern. It is therefore useful to refocus on risk management strategies available to the average farmer.

Risk Management Strategies

It is pertinent to understand strategies adopted by farmers to deal with risk. Risk management strategies refer to all actions and arrangements that are aimed at mitigating, transferring and coping with risks. Risk mitigation arrangements are aimed at preventing risky events from occurring and in the event of its occurrence, limit and cushion the effect of the severity of the attendant losses. Examples include extension services, integrated pest management practices etc. Risk transfer strategies are defined by arrangements that transfer risk to a willing third party, for a fee usually referred to as premium. A typical example is the agricultural insurance schemes of governments. Risk coping strategies refer to arrangements that enable a person or group of persons who have the misfortune of experiencing a risky event to cope with the attendant losses. This section will focus on two broad areas of risk management strategies namely informal and formal risk management strategies.

Informal Risk Management Strategies

Informal risk management strategies refer to all arrangements that do not come under the regulatory and supervisory mandate of government agencies and are aimed at mitigating, transferring and coping with risks. The arrangement may involve individuals or households groups of individuals. Examples of informal risk management strategy are enterprise diversification and application of new technologies.

Enterprise Diversification

Enterprise diversification as an informal risk management strategy is aimed at reducing income variability on the farm .This involves combining different production activities. It can take the form of mixed cropping, mixed farming or a combination of both. Enterprise diversification as a risk management tool hinges on the doctrine of not putting all ones eggs in one basket. This way, the chance of huge loss on account of a single mishap is appreciably reduced. Enterprise diversification is anchored on the assumption that not all the enterprises are destined to fail in a production season. According to Lee *et al* (1980), since each enterprise in an enterprise diversification scheme is a part of the total production process, a major loss in one is small in comparison to the total. It should be emphasized that for enterprise diversification to be effective, the included enterprises should not be subject to the same hazards.

Application of New Technologies

Development and Adoption of New and Improved production methods can be seen as risk management strategy.

Improved technological packages such as improved seeds and precision farming can be employed in managing risks in agricultural production. According to NAIC (2017), some improved seed varieties can be resistant to weeds, pest and diseases. Precision farming as a technology package and risk management strategy seeks to control the rate of planting seeds and the application of fertilizer and pesticide in the fields on a per hectare basis. This strategy is as opposed to the conventional approach which applies the same rate across the entire crop field. It should be underlined that application of new technologies can be both a risk management strategy as well as a source of risk and uncertainty. For instance technological uncertainty can come about through the rapidity of technological change. So, success or failure in the application of new technologies as risk management strategy is dependent on the adoption of the recommended packages. Furthermore, reducing the effect of new technologies as a source of risk and uncertainty is partly dependent on the ability of the farmer to change with the changing technology.

It should be noted that application of new technologies can be categorized as both informal and formal risk management strategies. The process of generation and dissemination of the new technologies can come under both institutional and non-institutional mandates. The adoption process is obviously an informal process.

Formal Risk Management Strategies

Formal risk management strategies refer to all arrangements that come under the regulatory and supervisory mandate of government agencies and are aimed at mitigating, transferring and coping with risks. Formal risk management strategies are market-based activities and are publicly provided. Examples of formal risk management strategies include contract farming and agricultural insurance.

Contract Farming

According to Eaton and Shepherd (2001), contract farming can be defined as an agreement and arrangement between farmers and processing and/or marketing firms for the production and supply of agricultural products under forward agreements, usually at predetermined prices. The agreement and arrangement frequently entails the specification that the purchaser provides particular production support by way of the supply of farm inputs and extension services. The arrangement also provides for a commitment on the part of the farmer to provide a named commodity in quantities and at qualities determined by the purchaser. A contract farming arrangement further provides a commitment on the part of the company to support the farmer's production and to purchase the commodity. Lee *et al* (1980), stated that contract farming is an avenue for a farm operator to obtain funds for the enterprise and went further to identify three basic types of contract farming namely market specification contract, production management contracts and resource-providing contracts.

Market specification contract refers to an agreement in which farm inputs or outputs are exchanged at some stated future date for an agreed price. Production management contracts, according to Lee *et al* (1980), has the same features as market specification contract in addition to the farmer receiving technical advice and management services from the input supplier or processor. Resource-providing contract enables the farmer receive financing and guaranteed management market outlet and production management assistance from the marketing firm.

One of the strong points for contract farming is that instead of being solely borne by the farmer, most of the marketing and production risks are shared by the contacting agency and the contracted farmer. Because of the provision of specifying prices in advance, contract farming agreement can reduce risk and uncertainty. However, if and when the contracts are not based on fixed prices but are rather based on market prices at the time of delivery then the contracted farmer is clearly dependent on market volatility. It is therefore self evident that the extent to which contract farming mitigates market risk depends on the terms of the contract agreement. Eaton and Shepherd (2001), cautions that farmers entering new contract farming ventures should be essentially balance the prospect of higher returns with the possibility of greater risk.

AGRICULTURAL INSURANCE AS A RISK MANAGEMENT STRATEGY

Agricultural insurance is a risk management strategy. Formal agricultural insurance is a formal risk management strategy. Formal agricultural insurance is not as efficient as is expected in Nigeria; hence formal agricultural insurance services in Nigeria are more easily available to wealthier, larger and more well-established farmers. This next section is devoted to focusing on the role of agricultural insurance as a risk management strategy.

Agricultural insurance is veritable financial tool with which to reduce the adverse effects of the risks inherent in agricultural production. With an agricultural insurance policy, risks are transferred to a third party through an insurance contract. The risk reducing ability of agricultural insurance in fact transcends agricultural production and encompasses agricultural processing, marketing and the entire value chain. Agricultural insurance serves the basic purpose of providing protection against insurable risk arising from adverse events. Insurable risk refers to the eventuality for loss or damage that is definable, fortuitous, and similar to a large number of known exposures and pays premium that is commensurate with the potential loss (Online Business Dictionary - BusinessDictionary.com, n.d.). In the view of Lee *et al.* (1980), an event is insurable if the probability of its occurrence can be predicted and the cost of the event to be insured party can be determined. Thus the event must be important enough to cause appreciable economic hardship to the insurance holder. According to Lee *et al.* (1980), the rudiment for all insurance is the law of large numbers. This law states that actual results tend to equal expected results as the number of cases increases. For instance, a mishap for an individual whose agricultural enterprise is covered by an agricultural insurance policy becomes in the pool, an easily borne, relatively constant, annual loss for the relevant insurance company.

If agricultural risks are not properly managed, risky events when they occur, as they do quite often, tend to increase the variability of farm income and to that extent sufficiently detract from the profitability of the agricultural enterprise. Let us focus on the role of agricultural insurance in risk management strategy in Nigeria with special emphasis on the Nigerian Agricultural Insurance Corporation (NAIC).

The Nigerian Agricultural Insurance Scheme was established on 15th of November 1987. Implementation of the Scheme was originally vested in the Nigerian Agricultural Insurance Company Limited. This company was later incorporated in June, 1988 and eventually turned into a Corporation in 1993 by the enabling Act 37 of 1993.Motivation for the establishment was based on unwillingness of conventional insurance companies to accept agricultural risks which are considered rather too risky.

The following agricultural insurance services are available to farmers under the auspices of the Nigerian Insurance Corporation: commercial crop insurance, livestock insurance, and other general services. Commercial crop insurance refers to non-food crops of economic value grown sole and on a large scale. The insurance holders pay 100% premium as there is no government subsidy for this category. Some examples of crops in this category include cotton, pineapple, citrus, rubber, plantain etc. Subsidized livestock insurance policy is an insurance cover for livestock. Subsidized livestock insurance policy holders pay 50% of the premium and the remaining 50% is borne by the Federal and State Governments of the location where the farm is located in a ratio of 3:1. Perils undercover include death due to disease, accident, fire, lightening, storm and flood. Motor insurance is the most common form the general insurance category. It is designed to protect insured motor vehicles for loss or damage to vehicles.

Other contributions of the NAIC to risk management in the agricultural value chain include covering the risks involved in the production, importation and distribution of fertilizers, seeds and seedlings and other agricultural inputs. NAIC also provides insurance covers which are provided in any part of Nigeria where the storage and processing infrastructures are located. NAIC does this using the fire/special perils, burglary/theft and multiple perils insurance policies. Furthermore, NAIC links its insurance products to the loans provided by deposit money banks and other financial institutions to investors engaged in the production, storage, Processing and marketing of agricultural commodities in Nigeria. According to NAIC (2017), the relevant policies cover the risks associated with weather and other non-weather perils. NAIC linkage with the banks and other financial institutions has turned out to be a constant requirement for farmers seeking to obtain from formal sources to compulsorily obtain insurance covers for their enterprises. However, for most farmers, obtaining compensation after a mishap tends to be an uphill task.

It is evident that the NAIC tends not to cover the agricultural production activities of the smallholder farmers who are incidentally in the majority of agricultural production in Nigeria. This may be a pointer to the neglect of smallholder agriculture. It may also be on account of the poverty-linked characteristics of the smallholder farmers whose savings are too low to sustain payment of the insurance premiums. It therefore stands to reason that the agricultural insurance services easily available to the smallholder farmers should be in the areas of enterprise diversification, contract farming, government subsidy etc.

SOLUTIONS AND RECOMMENDATIONS

Risk and uncertainty cannot be totally eliminated in an enterprise. The import of this statement and the need to cover the attendant risks become more pertinent in an enterprise that is as inherently risky as agriculture. Government price support programmes can help reduce price. Financial risk can be reduced by ensuring that enterprises have strong equity base. This is because debt capital costs have a direct and positive relationship with financial risk.

In the views of Agriculture for Impact (2012), agriculture by its nature is a private sector activity. This is so to the extent almost all the enterprises that produce agricultural goods and services are owned and managed by such private entities and units as firms and households. As a result, the private sector in Nigeria in general and the farm households in particular have a lot to offer by way of investments in informal risk management strategies.

The idea of agriculture being a private sector activity notwithstanding, there are fundamental justifications for public-sector intervention in the agricultural sector. Arguments in support for public-sector involvement in the economy include the market failure argument and the inequality or poverty argument (Olomola *et. al.* 2014). When the competitive market system either produces the "wrong" amounts of goods and services or fails to allocate any resources whatsoever to the production of some economically justified goods and services, market failure is implicated (McConnel and Brue, 1999). According to Oji (2002), a market failure can arise from imperfect information available to producers and consumers of goods and services; he concluded that governmental action or policy can correct all types of market failures. Government intervention is therefore needed by way of investment in formal risk management strategies. For instance, it has been stated that contract farming mitigates market risk and in turn depends on the terms of the contract agreement. Government intervention in the area of contract farming can come by way of ensuring contract enforcement so as to give meaning and effect to the risk management strategy.

Relevant institutions involved in agricultural insurance as a formal risk management strategy should ensure that famers obtain due compensation for their insured enterprises in the event of the relevant catastrophe in the farm enterprise. This is a veritable area for investment by government. When farmers have the institutional backing that the premiums paid on account of taking an insurance policy count and that in the event of loss they, the farmers, will duly receive indemnity to the degree of the loss, then a solution would have been found for one of the problems of agricultural risk management services in Nigeria.

Agricultural insurance should essentially be made to cover all farm enterprises including small-scale farmer enterprises. Incentivizing and subsidizing the practice of agriculture, including small-scale agriculture, improves incomes and strengthens livelihoods of the majority of the people. To all intents and purposes, this should be the concern of a responsible government. According to FAO, (2012), small-holder farmers usually face various forms of poverty such as weak property rights, poor access to markets and financial services, and severely limited ability to manage risks. It therefore stands to reason that providing agricultural insurance services to all farm enterprises by way government incentivizing and subsidizing the practice of agriculture should be guaranteed.

FUTURE RESEARCH DIRECTIONS

There is the need for constant re-examination of the role agricultural insurance as a risk management strategy. There is also the need for investigation into the efficiency of government's investment of tax payers' money in agricultural insurance services. There is therefore the obvious need for research focus on the efficiency and the actual beneficiaries of government's investment in incentives and subsidies by way of agricultural insurance schemes.

CONCLUSION

Agricultural insurance services can be provided to farmers by both formal and informal means. Neither formal nor informal agricultural insurance services are adequate on their own. It is important to note that efficient formal agricultural insurance services are more suited to enable farmers manage some perils. Good on-farm risk-management practices and sound production and farm management techniques should essentially complement the formal agricultural insurance practices. To all intents and purposes, agricultural insurance cannot be regarded as a universal solution to the risk and uncertainties that confront farmers. It can only indemnify a part of the losses resulting from some of the perils. Granted that the practice of agriculture is a private sector activity, the need for institutional support especially in the area formal agricultural insurance schemes cannot be overemphasized.

REFERENCES

Bosede, A. J. (2010). Economic assessment of fertilizer use and integrated practices for environmental sustainability and agricultural productivity in Sudan savannah zone, Nigeria. *African Journal of Agricultural Research*, *5*(5), 338–343.

Food and Agricultural Organization (FAO). (2012). *The state of food and agriculture*. Rome: FAO.

Guerron-Quintana, P. A. (2012). *Risk and Uncertainty*. Retrieved from www.philadelphiafed.org/ research-and-data/publications

Horne, J. V. (1980). *Financial Management and Policy*. Prentice/Hall International London.

Knight, F. H. (1965). *Risk, Uncertainty, and Profit*. New York: Harper and Row.

Langemeier, M. (2015). *What is risk and uncertainty?* Center for Commercial Agriculture, Purdue University.

Lee, W.F., Boehlje, M.D., Nelson, A.G. & Murray, W.G. (1980). *Agricultural finance* (7th ed.). Ames, IA: The Iowa State University.

McConnel, C. R., & Breu, S. L. (1999). Macroeconomics: Principles, Problems and Polices (14th ed.). McGraw Hill Coy Inc.

Nigerian Agricultural Insurance Corporation (NAIC). (2017). *Insurable risks in the agricultural value chain.* Retrieved from www.naic.gov.ng

Nto, P. O., Mbanasor, J. A., & Nwaru, J. C. (2011). Analysis of risk among agribusiness enterprises in Abia State Nigeria. *Journal of Economics and International Finance*, *3*, 187–194.

Oji, K. O. (2002). *Basic Principles of Economics for Agricultural Projects and Policy Analyses*. Nsukka, Nigeria: Prize Publishers.

Online Business Dictionary - BusinessDictionary.com. (n.d.). Retrieved March 23, 2017, from http://www.businessdictionary.com/

Surbhi, S. (2016). *Difference between risk and uncertainty*. Retrieved from http//keydifferences.com/author/surbhi

Teweldemedhin, M.Y., & Kapimbi, Y. (2012). Factors influencing enterprises diversification as a risk strategy management in Namibia; A case study of communal farmers from the Kunene region. *International Journal of Agricultural Sciences*, *2*(9), 845–853.

Index

G

I

L

M

N

P

R

S

U

V

www.ingramcontent.com/pod-product-compliance
Lightning Source LLC
LaVergne TN
LVHW081323300926
849161LV00008B/1859
* 9 7 8 1 5 2 2 5 3 0 5 9 6 *